DERMATOLOGY - LABORATORY AND CLINICAL RESEARCH

SUN EXPOSURE

RISK FACTORS, PROTECTION PRACTICES AND HEALTH EFFECTS

DERMATOLOGY - LABORATORY AND CLINICAL RESEARCH

Additional books in this series can be found on Nova's website under the Series tab.

Additional e-books in this series can be found on Nova's website under the e-book tab.

DERMATOLOGY - LABORATORY AND CLINICAL RESEARCH

SUN EXPOSURE

RISK FACTORS, PROTECTION PRACTICES AND HEALTH EFFECTS

COLLIN P. RAINES
EDITOR

NOVA BIOMEDICAL

New York

Library of Congress Cataloging-in-Publication Data

ISBN: 978-1-63482-087-5

Library of Congress Control Number: 2015930887

Published by Nova Science Publishers, Inc. † New York

Contents

Preface

Sun protection practices play a prominent role in dermatology in terms of photoaging, photocarcinogensis, and photodermatoses. This book begins by outlining the various methods of sun protection. It continues to discuss risk factors for sun exposure during spring break among college students; the role sun exposure plays in skin aging; and protection habits and vitamin D levels in children and adolescents with a history of malignancy.

Chapter I – The ultraviolet (UV) spectrum of the solar light is the most damaging exogenous source for our skin. Photoaging affects the sun-exposed areas and is characterized clinically by fine and coarse wrinkling, roughness, dryness, laxity, telangiectasia, loss of tensile strength and pigmentary changes. There is also an increase in development of benign and malignant neoplasms on photoaged skin. UV radiation (UVR) penetrates our skin, reaches the cells and is absorbed by DNA, leading to the formation of photoproducts that inactivate the functions of DNA. UVA radiation acts mostly indirectly through the generation of Reactive Oxygen Species (ROS), producing high amounts of singled oxygen, which can further initiate lipid peroxidation, oxidation of proteins or generation of DNA strand breaks. UVB action is mostly by direct interaction with DNA via the induction of DNA damage. The epidermis and dermis are both affected by UVB, but the dermis is also affected to a significant extent by UVA. It has long been thought that the majority of human photo-lesions are due to UVB rays; now it is believed that UVA plays a substantial role in photoaging. But DNA is not the only biomolecule damaged by UVR. Free radicals and oxidants produced by UVR oxidize also lipids and proteins in the cell. The skin then springs into action with an inflammatory response, characterized by erythema (sunburn), the release of proteases and cytokines. Infectious agents that may try to take advantage of this compromised situation are sought out and destroyed. Then, a temporary period of immune suppression follows. During that time, repair systems are activated that excise DNA lesions and replace the damaged DNA, and non-functional proteins are turned-over by proteases. Hormesis effect activates the synthesis of melanin and antioxidant protection, and damaged lipids are cleaved and replaced. Irreparable cells are removed by apoptosis. However, these repair mechanisms are not 100% effective. The problem arises in the cases of intensive acute sun exposure or in the cases of chronic sun exposure over longer decades, which manifests as skin photoaging. To what extent each mechanism—mitochondrialDNA mutagenesis, protein oxidation, downregulation of collagen synthesis and increased expression of matrix metalloproteinases—contributes to premature skin aging is still not answered. UV protection includes not only reduction of sun exposure but also the use of sun protective filters, UV protective clothes, DNA repair enzymes, and antioxidant supplementation.

Chapter II – Sun protection practices play a prominent role in dermatology in terms of photoaging, photocarcinogensis, and photodermatoses. In this chapter, the author's outline the various methods of sun protection. Naturally occurring protective agents, such as geographic and environmental variations create a substantial difference in exposure to ultraviolet (UV) radiation. Physical protective barriers including glass, clothing, and sunglasses can provide UV protection, but there is great variability in their defensive properties. Sunscreens play a major role in sun protection practices and vary substantially in chemical composition and effectiveness regarding protection against UVA versus UVB radiation. The author's will discuss inorganic products versus organic products in sunscreens as well as their associated advantages and disadvantages. Finally, systemic agents, such as plant extracts, carotenoids, polyphenols, and other antioxidants may prove to be promising adjuncts for sun protection in the future.

Chapter III – Sun exposure, the main source of ultraviolet radiation exposure, is the major environmental risk factor for skin cancers, both melanoma and non-melanoma. The impact of exposure can be dramatically reduced by practicing sun protection behaviors. Still, recommendations for sunlight exposure are an area of controversy, as sun avoidance and sun protection practices may lead to inadequate vitamin D levels. This duality is particularly relevant to childhood cancer survivors, who are particularly vulnerable both to the hazardous effects of ultraviolet radiation and to deleterious skeletal and extra-skeletal effects of vitamin D deficiency. These patients are at high risk for developing non-melanoma skin cancer and therefore are firmly advised to avoid or minimize sun exposure and adopt skin protection measures. On the other hand, several studies showed an inverse association between regional UV-B radiation exposure and cancer mortality, with activation of vitamin D-related pathways by sun exposure proposed as the likely mechanism. A protective role for vitamin D in the prevention of and recovery from malignancies has been suggested by observational studies, by clinical trials showing reduced cancer incidence with vitamin D supplementation, and by laboratory studies demonstrating that the physiologically active form of vitamin D, 1;25$(OH)_2D_3$ has anti-cancerous effects. In this respect, it is worrisome that a number of studies demonstrated sub-optimal 25OHD levels in children and adolescents with a history of malignancy. Regarding sun habits of this population, there is very limited information in the literature comparing sun protection behaviors of childhood cancer survivors to individuals without a history of cancer. Three studies in adult survivors of childhood cancer showed inconsistent results. Only three previous studies investigated this issue in children and adolescents, showing that patients spent less time outside compared to controls, however adherence to sun protection recommendations was incomplete. Similar findings were recently observed by our group. The author's conclude that more attention should be paid to improve sun protection habits of survivors of childhood cancer throughout their lives. At the same time, since sunlight avoidance may result in vitamin D insufficiency and might have deleterious implications for bone health and possibly on cancer survival, dietary interventions to optimize intake of vitamin D may be required.

Chapter IV – In order to look at college students' behavioral practices prior to a sunny vacation (during spring break) along with their beliefs and attitudes, the author's recruited sororities and fraternities in the Midwestern USA to complete a self-administered questionnaire. Sorority and fraternity students were expected to have high UVR exposure due to a strong desire to tan. The questionnaire included information on sun exposure during spring break, sun-sensitivity, and tanning attitudes and behaviors. Analyses examined associations between potential risk factors for spending 16 or more hours in the sun during spring break using logistic regression while controlling for the clustering effects of sororities and fraternities. Students who tanned mildly were 1.6 times more likely than those with moderate or deep tans to spend 16+ hours in the sun during spring break, suggesting a strong desire to tan. Students who spent 16+ hours in the sun during spring break were more likely to have frequented tanning beds (odds ratio of 2.4 for 11+ times vs. $\leq$5 times) and to have used self-tanning creams (odds ratio of 2.9) between New Years and spring break. These data provide evidence that use of artificial tanning devices and self-tanning creams or sprays among college students are related to increased intermittent sun exposure (during a spring break vacation) rather than reduced exposure. Mistaken beliefs regarding a base tan as potentially beneficial need to be addressed by excellent science examining the base tan theory and translated to the public. Replacement of tanning bed use with safer sunless tanning creams may reduce some of the harmful UVR exposures. Education alone will not be sufficient to change sun seeking behavior as was seen here and in other studies.

In: Sun Exposure
Editor: Collin P. Raines
ISBN: 978-1-63482-087-5

Chapter I

The Role of Sun Exposure in Skin Aging

Raja Dahmane, MD, PhD[1,2,*], Ruza Pandel, PhD[1,2], Polonca Trebse, PhD[1,2] and Borut Poljsak, PhD[1]

[1]Laboratory for Oxidative Stress Research, Faculty of Health Sciences, University of Ljubljana, Ljubljana, Slovenia
[2]Chair of Biomedicine in Health Care Division, Faculty of Health Sciences, University of Ljubljana, Ljubljana, Slovenia

Abstract

The ultraviolet (UV) spectrum of the solar light is the most damaging exogenous source for our skin. Photoaging affects the sun-exposed areas and is characterized clinically by fine and coarse wrinkling, roughness, dryness, laxity, telangiectasia, loss of tensile strength and pigmentary changes. There is also an increase in development of benign and malignant neoplasms on photoaged skin.

UV radiation (UVR) penetrates our skin, reaches the cells and is absorbed by DNA, leading to the formation of photoproducts that inactivate the functions of DNA. UVA radiation acts mostly indirectly

[*] Corresponding author: Raja Dahmane, MD, PhD, Faculty of Health Sciences, University of Ljubljana, Zdravstvena pot 5, 1000 Ljubljana, Slovenia. Phone: + 386 40900878, fax: +386 1 3001119, e-mail: raja.dahmane@guest.arnes.si.

through the generation of Reactive Oxygen Species (ROS), producing high amounts of singled oxygen, which can further initiate lipid peroxidation, oxidation of proteins or generation of DNA strand breaks. UVB action is mostly by direct interaction with DNA via the induction of DNA damage. The epidermis and dermis are both affected by UVB, but the dermis is also affected to a significant extent by UVA. It has long been thought that the majority of human photo-lesions are due to UVB rays; now it is believed that UVA plays a substantial role in photoaging. But DNA is not the only biomolecule damaged by UVR. Free radicals and oxidants produced by UVR oxidize also lipids and proteins in the cell. The skin then springs into action with an inflammatory response, characterized by erythema (sunburn), the release of proteases and cytokines. Infectious agents that may try to take advantage of this compromised situation are sought out and destroyed. Then, a temporary period of immune suppression follows. During that time, repair systems are activated that excise DNA lesions and replace the damaged DNA, and non-functional proteins are turned-over by proteases. Hormesis effect activates the synthesis of melanin and antioxidant protection, and damaged lipids are cleaved and replaced. Irreparable cells are removed by apoptosis. However, these repair mechanisms are not 100% effective. The problem arises in the cases of intensive acute sun exposure or in the cases of chronic sun exposure over longer decades, which manifests as skin photoaging.

To what extent each mechanism—mitochondrialDNA mutagenesis, protein oxidation, downregulation of collagen synthesis and increased expression of matrix metalloproteinases—contributes to premature skin aging is still not answered.

UV protection includes not only reduction of sun exposure but also the use of sun protective filters, UV protective clothes, DNA repair enzymes, and antioxidant supplementation.

Introduction

Human skin, like all other organs, undergoes chronological and biological aging. Aging of the skin is a composite of actinic damage, chronological aging, and internal influences. In addition, unlike other organs, skin is in direct contact with the environment and, therefore, undergoes aging as a consequence of environmental damage (Fisher et al., 2002). Factors contributing to premature aging are dependent on age, sex, pigmentation, smoking, sun exposure history, alcohol consumption and other environmental, and lifestyle factors (Ernster et al., 1995). Age-related physiological changes in elderly skin include clinical, histological, and biochemical changes, as well as changes in neurosensory perception, barrier function, wound healing and higher incidence of benign and cancerous diseases (Rasche and Elsner, 2010). Skin aging appears to be the result of two types of aging, intrinsic and extrinsic aging. Extrinsic aging is the skin's response to external damage and is controllable to a very large degree by the lifestyle choices we make every day. The rate of aging is significantly different among different populations, as well as among different anatomical sites, even within a single individual. The intrinsic rate of skin aging in any individual can also be dramatically influenced by personal and environmental factors, particularly the amount of exposure to ultraviolet (UV) light. Photodamage, which considerably accelerates the visible aging of skin, also greatly increases the risk of cutaneous neoplasm (Farage et al., 2008). Damage to human skin due to UV light from the sun (photoaging) and damage occurring as a consequence of the passage of time (chronologic or natural aging) were considered to be distinct entities. The findings of the study performed by Varani et al. (2000) indicate that naturally aged sun-protected skin and photoaged skin share important molecular features, including connective tissue damage, elevated matrix metalloproteinase levels, and reduced collagen production. The intrinsic (genetically determined) and the extrinsic (UV and toxic exposure mediated) skin-aging processes are thus overlapped and are strongly related to the increased generation of free radicals in the skin. Oxidative stress is believed to underlie changes associated with both photoaging and natural aging and is considered of primary importance in driving the skin-aging process. Extrinsic skin aging develops due to several factors: ionizing radiation, severe physical and psychological stress, alcohol intake, poor nutrition, overeating, environmental pollution, and exposure to UV radiation. It is estimated that among all these environmental factors, UV radiation contributes up to 80%. UV radiation is the most important environmental factor in the development of skin cancer and skin aging. After age 45, a thinning of the skin begins, due in part to hormonal changes. This thinning makes the skin more fragile and vulnerable to damage by abrasion and more sensitive to irritating environmental factors and allergens. Thin, wrinkled skin is very often attributed to a lack of collagen. The dermis and overlying epidermis of aging skin are profoundly altered (Schmid et al., 2002). Slower protein synthesis is one of the most common events observed during aging. The synthesis of both structural proteins, such as collagen, and enzymes that repair and maintain the normal metabolic functions of the cell, is slowed down. This leads to the inefficient removal of damaged molecules and decreased intra- and intercellular signaling pathways (Rocquet and Bonté, 2002). Synthesis of types I and III procollagen is reduced in aged skin (Varani et al., 2000).

There is a progressive disappearance of elastic tissue in the dermis due to reduction in elastin gene expression after the age of 40 to 50. Additionally, decreased proliferative capacity of skin cells and decreased matrix synthesis contribute to intrinsic skin-aging process. The moisture-holding proteoglycans and glycosaminoglycans decrease in abundance, making the skin become dryer and looser. Proteoglycans make up a major part of the extracellular matrix, the material between cells that provides structural support. Proteoglycans are heavily glycosylated glycoproteins. This means that they are proteins with chains of polysaccharides, a kind of carbohydrate, attached. The skin loses fat, so it looks less plump and smooth. The number of blood vessels in aged skin decreases and the skin loses its youthful color and glow. Since blood circulation in the dermic layer slows down, the delivery of nutrients and oxygen to skin cells is decreased.

Skin atrophy is marked only after the fifth decade of human life and shows a plethora of histomorphologic changes including epidermal thinning, flattening of the dermal-epidermal junction, loss of melanocytes, and immune-competent Langerhans cells (Bhattacharyya, 2010). There are also dermal changes such as reduced fibroblast population and sebaceous glands. There are typical ultrastructural changes in microvasculature of elderly people. Two of the most noticeable changes as skin ages are alterations to pigment production (e.g., age spots) and the formation of wrinkles.

Altered melanocyte function and reorganized, cross-linked highly structured collagen matrices directly drive these visible aging changes respectively, but the primary cause of both is excessive lifetime exposure to the sunlight (Jenkins et al., 2009). The major histological features of sun-protected, intrinsically aged skin include a thin epidermis, significant flattening of the dermal-epidermal junction (this results in a reduced exchange of nutrients and metabolites between these two parts), thinning of the dermis and subcutaneous adipose layer and reduced numbers of keratinocytes, fibroblasts, Langerhans cells, mast cells and melanocytes. The dermis appears hypocellular with fewer fibroblasts and mast cells and loss of dermal volume.

There is a decrease in the number of dermal blood vessels and a decrease in the density of Pacinian and Meissner's corpuscles, responsible for pressure and light touch perception (Yaar and Gilchrest, 2003).

Besides chronological aging, actinic aging, also called photodamage, causes premature skin aging: thinning of the dermis, a loss of collagen content and protein organization (Bolognia, 1993). There is also a depressed sensory and autonomic innervation of epidermis and dermis.

Mechanisms of Skin Photoaging

Because UV and photoaging play very important roles in skin aging, this chapter describes their damaging effects on skin in a more detailed way. Unlike chronological aging, which depends on the passage of time per se, photoaging depends primarily on the degree of sun exposure and skin pigment. Individuals who have outdoor lifestyles, live in sunny climates, and are lightly pigmented will experience the greatest degree of UVR skin penetration and thus suffer from the effects of the photoaging (Fisher et al., 2002).

The term "photoaging" (also known as "Dermatoheliosis") was first coined in 1986, and describes the effects of chronic UV light exposure on skin (Kligman and Kligman, 1986). Photoaging refers to the physiologic and pathological changes that occur specifically in aged tissue that has experienced chronic sun exposure over time. Human skin aging resulting from UV irradiation is a cumulative process that occurs based on the degree of sun exposure and the level of skin pigment. Clinical signs of photoaging include wrinkles; mottled pigmentation; rough skin, and loss of skin tone; dryness; irregular, dark/light pigmentation; sallowness; either deep furrows or severe atrophy; telangiectasia; premalignant lesions; laxity; and a leathery appearance (Yaar et al., 2002). Other signs include elastosis (a coarse, yellow, cobblestoned effect of the skin) and actinic purpura (easy bruising related to vascular wall fragility in the dermis) (Gilchrest, 1990). Sun-exposed areas of the skin, such as the face, neck, upper chest, hands, and forearms, are the sites where these changes occur most often (Helfrich et al., 2008). While intrinsically aged skin does not show vascular damage, photodamaged skin does. Studies in humans and in the albino and hairless mouse model for skin aging have shown that acute and chronic UVB irradiation greatly increases skin vascularization and angiogenesis (Yano et al., 2002).

Skin Exposure to UVR

The sun is the main source of UVR and the main contributor to the photoaging. In order to understand sun's effects on the skin, a brief introduction on basic characteristics of UV light and its environmental exposure will be presented. Solar radiation reaching the earth's surface includes wavelengths in the range 290 to 4000 nm and is divided into three bands: UV radiation (290 to 400 nm), visible light (400 to 760 nm) and IR (760 to 4000 nm). UVR (100 to 400 nm) comprises only 5% of the terrestrial solar radiation. Sun's UVR is divided into categories based on the wavelength.

- UVC Radiation - 100 to 290 nm. UVC radiation is almost completely absorbed by the ozone layer and does not affect the skin. UVC radiation can be found in artificial sources such as mercury arc lamps and germicidal lamps.
- UVB Radiation - 290 to 320 nm. UVB affects the outer layer of skin, the epidermis, and is the primary agent responsible for sunburns. It is the most intense between the hours of 10:00 am and 2:00 pm, when the sunlight is brightest. It is also more intense in the summer months, accounting for 70% of a person's yearly UVB dose. UVB does not penetrate glass.
- UVA Radiation - 320 to 400 nm. UVA was once thought to have a minor effect on skin damage, but now studies are showing that UVA is a major contributor to skin damage. UVA penetrates deeper into the skin and works more efficiently. The intensity of UVA radiation is more constant than UVB without the variations during the day and throughout the year. UVA is not filtered by glass.

According to Laurent-Applegate and Schwarzkopf (2001), "the UVA has greater penetration (e.g., about 20% at 365 nm). Whereas UVB is much more damaging to the skin than UVA if equal exposures are carried out; the deeper penetration of UVA and its greater abundance in sunlight (about 95% UVA, 5% UVB) suggest that it can also be a significant contributor to damage. As the photons in the UVA waveband are less energetic, significantly more photons are needed to cause the same damage as that induced by the shorter wavelengths in the UVB region. It is important to remember that UVA photons are present in sunlight in much higher quantities than those of UVB and that these longer wavelengths have the potential to penetrate into the dermis to a far greater extent than UVB because of its less energetic potential" (Laurent-Applegate and Schwarzkopf, 2001).

In the middle age, it was known the damaging effect of sun, which caused "farmer's skin," and pale skin was appreciated. In the last decades, this trend changed. Dark skin complexion was propagated as a sign of "healthy skin."

The exposure of human skin to environmental and artificial ultraviolet irradiation has increased significantly in the last 50 years. This is not only due to an increased solar UV irradiation as a consequence of the stratospheric ozone depletion, but also the result of an inappropriate social behavior with the use of tanning parlors being very popular. Besides this, leisure activities and living style with travelling to equatorial regions also add to the individual annual UV load. Total UVR load depends on time of exposure, duration and intensity of exposure.

Since the population in industrialized countries shows an increasing total lifespan, in parallel the cumulative lifetime dose of solar and artificial UV irradiation is dramatically augmented (Grether-Beck et al., 2005).

While there is no standard measure, sun exposure can be generally classified as intermittent or chronic, and the effects may be considered acute or cumulative. Intermittent sun exposure is obtained sporadically, usually during recreational activities, and particularly by indoor workers who have only weekends or vacations to be outdoors and whose skin has not adapted to the sun. Chronic sun exposure is incurred by consistent, repetitive sun exposure during outdoor work or recreation. Acute sun exposure is obtained over a short time period on skin that has not adapted to the sun (National Cancer Institute). UV-induced extrinsic aging is visible on chronically UV-exposed skin areas in persons frequently engaged in outdoor activities. Exposed skin surface is irradiated differently depending on cultural and social behavior, clothing, the position of the sun in the sky and the relative position of the body. Exposure to UVB of the most exposed skin surfaces, such as nose, tops of the ears and forehead, relative to that of the lesser exposed areas, such as underneath the chin, normally ranges over an order of magnitude. Ground reflectance plays a major role in exposure to UVB of the eyes and shaded skin surfaces, particularly with highly reflective surfaces such as snow (IARC, 2010). The cumulative annual exposure dose of solar UVR varies widely among individuals in a given population, depending to a large extent on occupation and extent of outdoor activities (IARC, 2010). For example, it has been estimated that indoor workers in mid-latitudes receive an annual exposure dose of solar UVR to the face of about 40 to 160 times "the minimal erythema dose" (MED), depending upon propensity for outdoor activities, whereas the annual solar exposure dose for outdoor workers is typically around 250 times the MED. Because few actual measurements have been reported of personal exposures, these estimates should be considered to be very approximate and subject to differences in cultural and social behavior, clothing, occupation and outdoor activities. Cumulative annual outdoor exposures may be augmented by exposures to artificial sources of UVR. For example, the use of cosmetic tanning appliances increased in popularity in the 1980s. The majority of users are young women, and the median annual exposure dose is probably 20 to 30 times the MED.

Currently, used appliances emit primarily UVA radiation; prior to the 1980s, tanning lamps emitted higher proportions of UVB and UVC (IARC, 2010).

Ozone Depletion and UVR

The quality and quantity of UVR at the earth's surface depend on the energy output of the sun and the transmission properties of the atmosphere. From a biological viewpoint, UVB radiation is by far the most significant part of the terrestrial UV spectrum, and the levels of radiation in this waveband reaching the surface of the earth are largely controlled by ozone, a gas which comprises approximately one molecule out of every two million in the atmosphere (Diffey, 1991).

Ozone layer in the stratosphere is also the major barrier to UVC (and largely to UVB). Depletion of this layer may result in more UVB reaching the earth's surface, with the corresponding increase in photochemical damage to living organisms. It has been estimated that each 5% depletion of stratospheric ozone will raise UVB flux at ground level by 10%. Up to 10% of UVB light falling on the skin can penetrate the epidermis to reach the dermis. In 1974, Molina and Rowland (Molina and Rowland, 1974) first warned that chlorofluorocarbons (CFCs) and other gases released by human activities could alter the natural balance of creative and destructive processes and lead to depletion of the stratospheric ozone layer. Substantial reductions of up to 50% in the ozone column observed in the austral spring over Antarctica and first reported in 1985 (Farman et al., 1985) are continuing (SORG, 1990).

Coupled with this, there has been a statistically significant downward trend in wintertime total ozone over the northern hemisphere of about 2% to 3% per decade for the past 30 years, although summertime ozone levels have remained approximately constant (Frederick, 1990).

In its report in June 1990, the UK Stratospheric Ozone Review Group concluded that there are serious limitations in our understanding and ability to quantify ozone depletion at the present levels of contaminant release and in our ability to predict the effects on stratospheric ozone of any further increases

(SORG, 1990). The atmosphere has a profound effect on the irradiance that reaches the surface of the earth.

In January (in the northern hemisphere) or July (in the southern hemisphere), when the solar elevation is low, direct UV travels a longer path through the atmosphere, and a large amount of scattering occurs.

In addition, much of the resultant scattered UV propagates downwards to the earth's surface at angles to the horizontal that are larger than the solar elevation, hence travelling a shorter and less absorptive path. This results in large ratios of scattered to direct UV. During the summer, the ratio of diffuse to direct UV is smaller (International program on chemical safety, Environmental Health Criteria, 160).

Factors Affecting Terrestrial UVR

As already mentioned, the spectral irradiance of UVR at the earth's surface is modified by temporal, geographical and meteorological factors (Frederick et al., 1989).

According to Diffey (1991) the following factors contribute significantly to terrestrial UVR intensity:

- *Time of day.* About 20% to 30% of total daily UVR is received one hour either side of midday in summer, with 75% between 9 am and 3 pm.
- *Season.* In temperate regions, the biologically damaging UVR reaching the earth's surface shows strong seasonal dependence. However, seasonal variation is much less nearer the equator.
- *Geographical latitude.* Annual UVR flux decreases with increasing distance from the equator.
- *Clouds.* Clouds reduce solar irradiance at the Earth's surface, although changes in the ultraviolet region are not as great as those of total intensity, since water in clouds attenuates solar infrared much more than UVR. The risk of overexposure may be increased under these conditions because the warning sensation of heat is diminished. Light clouds scattered over a blue sky make little difference to UVR intensity unless directly covering the sun, whilst complete light cloud cover reduces terrestrial UVR to about one-half of that from a clear sky. Even with heavy cloud cover, the scattered ultraviolet component

of sunlight (often called skylight) is seldom less than 10% of that under clear sky (Paltridge and Barton, 1978). However, very heavy storm clouds can virtually eliminate terrestrial UVR, even in summertime (Diffey, 1988).

- *Surface reflection.* Reflection of UVR from ground surfaces, including the sea, is normally low (<7%). However, gypsum sand reflects about 25% of incident UVB and fresh snow about 30% (Doda and Green, 1980, 1981) although other authors (Blumthaler and Ambach, 1985) have reported that the UVB reflectance of fresh snow exceeds 80%.
- *Altitude.* In general, each 1 km increase in altitude increases the ultraviolet flux by about 6% (Cutchis, 1980). Conversely, places on the earth's surface below sea level are relatively poorer in UVB content than nearby sites at sea level. This is strikingly apparent around the Dead Sea, 400 m below sea level (Kushelevsky and Slifkin, 1975).

UVR and Its Penetration to the Skin

UV incident on human skin can follow one of three courses: it can undergo absorption, reflection, or scattering. The first step in a photochemical reaction is the absorption of a single photon by a molecule and the production of an excited state in which one electron of the absorbing molecule is raised to a higher energy level. Such radiative transition can only occur efficiently when the photon energy of the radiation is close to the energy difference of the atom in the initial and final state (energy level).

The photochemistry that may then occur will, therefore, depend upon the molecular structure and the wavelength of UV as well as the specific reaction conditions. The primary products generated by UV absorption are generally ROS. Reflection not only occurs at the surface of the stratum corneum but at all interfaces changing in refractive index. Scattering occurs because of the different structural elements, such as hair follicles and sebaceous glands, and also by cellular components, such as mitochondria and ribosomes.

The remaining UV can penetrate into deeper skin layers (International program on chemical safety, Environmental Health Criteria, 160).

UV penetrates into the dermis exposing a variety of cells and structures, depending in part on the thickness of the human stratum corneum and epidermis. The depth of penetration is wavelength dependent—the longer the wavelength the deeper the penetration (Bruls et al., 1984).

Damaging Effect of UVR

Studies in hairless mice demonstrated the carcinogenicity of exposures to UVR in the wavelength ranges 315 to 400 nm (UVA), 280 to 315 nm (UVB) and ¾ 280 nm (UVC), UVB radiation being the most effective, followed by UVC and UVA. UVB radiation is three to four orders of magnitude more effective than UVA. Nevertheless, both short-wavelength UVA (315 to 340 nm) and long-wavelength UVA (340 to 400 nm) induced skin cancer in hairless mice. The carcinogenic effectiveness of the latter waveband is known only as an average value over the entire range; the uncertainty of this average is about one order of magnitude. In none of the experiments involving UVC was it possible to exclude completely a contribution of UVB, but the size of the effects observed indicate that they cannot be due to UVB alone (IARC, 2010). UVB is three to four times more effective than UVA in producing erythema. In humans, pigmentation protects against erythema and histopathological changes. People with a poor ability to tan, who burn easily and have light eye and hair color are at a higher risk of developing melanoma, basal-cell and squamous-cell carcinomas (IARC, 2010). UVB most commonly causes damage in the form of cyclobutane pyrimidine dimmers. UVA, on the other hand, primarily causes DNA damage indirectly by the production of short-lived Reactive Oxygen Species such as singlet oxygen, superoxide and H_2O_2 via endogenous photosensitizers. UVA radiation generates more phosphodiester bond breaks in DNA than would be expected by the total amount of energy directly absorbed by the DNA. Therefore, most likely there is indirect damage to DNA accomplished by endogenous photosensitizers such as riboflavin, nicotinamide coenzymes, and rare RNA bases (Laurent-Applegate and Schwarzkopf, 2001).

Skin Response to Acute and Chronic UV Radiation

Skin pigmentation is induced both by UVB and UVA rays. According to Diffey (1991), the normal responses of the skin to UVR can be classed under two headings: acute effects and chronic effects. An acute effect is one of rapid onset and generally of short duration, as opposed to a chronic effect, which is often of gradual onset and long duration. These effects should be distinguished from acute and chronic exposure conditions, which refer to the length of the UVR exposure. The acute reactions considered will be sunburn, tanning and vitamin D production.

Photoaging and skin cancer will be discussed as those chronic reactions produced by prolonged or repeated UVR exposure (Diffey, 1991). Exposure to ultraviolet (UV) radiation increases skin pigmentation and usually results in an even darkening of the skin.

However, it may also occasionally lead to the development of hyperpigmented lesions due to a local overproduction of pigment.

Skin Antioxidant Defense

Although the skin possesses an elaborate antioxidant defense system to deal with UV-induced oxidative stress and immunotoxicity, excessive and chronic exposure to UV light can overwhelm the cutaneous antioxidant and immune response capacity, leading to oxidative damage and immunotoxicity, premature skin aging, and skin cancer. Photoaged skin has significantly reduced concentrations of antioxidant enzymes in the stratum corneum and the epidermis, while the concentration of oxidized proteins in the upper dermis is increased. Acute exposure to UV irradiation depletes the catalase activity in the skin and increases protein oxidation (Sander et al., 2002).

DNA Damage

DNA is constantly exposed to DNA-damaging agents. Environmental DNA-damaging agents include UV light and ionizing radiation, as well as a variety of chemicals encountered in foodstuffs, or as air- and water-borne agents. Endogenous damaging agents include metabolites that can act as alkylating agents and the ROS that arise during respiration.

Since there are always less antioxidants available to neutralize all oxidizing agents involved in free radical formation and damage, a third line of defense evolved—damage repair mechanisms. Defense mechanisms are available to cope with constant free radical-induced damage in our cells. Among them are DNA repair systems to replace damaged DNA bases and antioxidants to neutralize free radicals. Damaged cells are eliminated by apoptosis and non-functional proteins are turned-over by proteases. But DNA repair capacity has been found to decrease with age.

For example, decrease in the level of proteins that participate in nucleotide excision repair was reported to occur for aged dermal fibroblasts (Goukassian et al., 2000).

Oxidation of DNA can produce different types of DNA damage: strand breaks, sister chromatid exchange, DNA-protein cross-links, sugar damage, abasic sites, and base modifications. Cell death, chromosome changes, mutation and morphological transformations are observed after UV exposure of prokaryotic and eukaryotic cells. Numerous types of UV-induced DNA damage have now been recognized that include stand breaks (single and double), cyclobutane-type pyrimidine dimers, 6-4 pyo photoproducts and the corresponding Dewar isomer, thymine glycols, 8-hydroxy guanine, and many more. Additionally, the specific lesions in DNA that can be induced by UVA radiation include pyrimidine dimmers, single-strand breaks (both not thought to be the critical lesions in UVA radiation-induced cellular lethality), and, perhaps more importantly, DNA protein cross-links (Peak et al., 1987; Peak et al., 1988). Solar UV causes formation of ROS, which can oxidize guanine in DNA to form 8-hydroxy-7,8-dihydroguanine (8-OHdG).

The frequency of this characteristic mutation in human skin increases with cumulative sun exposure and could be used as internal dosimeter of cumulative sun exposure (Yarosh, 2010). $OH^{\cdot}$ can add on to guanine at positions 4, 5 and 8 in the purine ring. Addition to C-8 produces a C-8 OH adduct radical that can be reduced to 8-hydroxy-7,8-dihydroguanine, oxidized to 8-hydroxyguanine, or undergo opening of the imidazole ring, followed by one-electron reduction and protonation, to give 2,6-diamino-4-hydroxy-5-formamidopyrimidine, abbreviated as FAPyG (Halliwell and Gutterigde, 1999). Photoexcitation of cytosine and guanine may lead to the formation in relatively minor yields of 6-hydroxy-5,6-dihydrocytosine and 8-oxo-7,8-dihydroguanine, respectively.

A second mechanism that requires the participation of endogenous photosensitizers together with oxygen is at the origin of most of the DNA damage generated by the UVA (320 to 400 nm) and visible light.

However, it may be expected that the latter oxidized purine lesion together with DNA strand breaks and pyrimidine base oxidation products are also generated with a lower efficiency through Fenton type reactions (Cadet et al., 1997). The number of different DNA modifications that $OH^{\cdot}$ is capable of producing appears to be over 100 (Hutchinson, 1985). In addition, DNA-protein cross-links are produced during UV exposure. Larger scale genetic alterations include chromosome breakage, sister chromatid exchanges and chromatid aberrations. Although partial UV action spectra are now available for many of these lesions, the most studied have been the different types of pyrimidine dimers (International program on chemical safety, Environmental Health Criteria, 160).

Damage to Elastin and Collagen and Wrinkle Formation

Wrinkling in the skin is caused by habitual photoaging, facial expressions, aging, smoking, poor hydration, and various other factors (Anderson, 2006). The effects of aging on the dermal layer are significant. Not only does the dermal layer thin but also less collagen is produced, and the elastin fibers that provide elasticity wear out. These changes in the scaffolding of the skin cause the skin to wrinkle and sag. Photoaged skin can be associated with either increased epidermal thickness or pronounced epidermal atrophy. The most pronounced histological change is the accumulation of elastin-containing material just below the dermal-epidermal junction, known as solar elastosis (Lavker, 1995). Collagen, which composes over 90% of the skin's total proteins, becomes disorganized (Bernstein et al., 1996). While elastin levels are increased in photoaged skin, levels of types I and III collagen precursors and cross-links are reduced (Talwar et al., 1995). It is likely that such changes in collagen precursors lead to reduced levels and/or altered organization of fibrillar collagen and thus may contribute to the wrinkled appearance of photodamaged human skin. The Glogau classification system was developed to objectively measure the severity of photoaging and especially wrinkles. Four types of skin depressions can be defined according to their depth: folds, permanent wrinkles, reducible wrinkles and skin micro-relief. When UV rays from the sun penetrate the skin, they damage elastic fibers. As the elastin weakens, skin becomes less elastic and loses its ability to snap back after being stretched. Photoaged skin contains elastic material in the reticular dermis, and the fibrillin deposits in the reticular dermis are enlarged. Elastic fibers have a central core of hydrophobic cross-linked elastin surrounded by fibrillin-rich microfibrils. The papillary dermal microfibrillin-rich microfibril network is truncated and depleted in photoaged skin.

There are fewer fibrillin-rich microfibrils in wrinkled photoaged skin, probably due to inflammatory cell proteinases (neutrophil elastase) or activation of matrix metalloproteinase (Tsuji et al., 2001). Cross-linking causing decreased elasticity could also be involved in wrinkle formation

(Watson et al., 1999). Wrinkles form especially in parts of skin that get stretched and move a lot (like around eyes, mouth and nose). Anti-aging skin care products should focus on wrinkles prevention also as anti-sun damage products in order to prevent elastin and collagen damage. A decrease in the overall ROS load by efficient sunscreens or other protective agents may represent promising strategies to prevent or at least minimize ROS-induced photoaging.

For example, the study of Seo et al. (2001) on photoaged skin has shown that UV irradiation increases the tropoelastin mRNA in keratinocytes and fibroblasts (Seo et al., 2001). Selective inhibition of skin fibroblast elastase could be one way to fight wrinkle formation following cumulative ultraviolet B irradiation (Tsukahara et al., 2001). Lysozyme may alter the elastic fibers in the surface, preventing further degradation and the accumulation of altered elastic fibers.

Why UV Radiation Accelerates the Aging Process?

As already mentioned, UV radiation can cause significant DNA damage and especially to mitochondrial DNA (mtDNA). In a study by Berneburg et al. (2004), it was confirmed that repetitive exposure to UVA light leads to an approximately 40% increase in the level of the common deletion in skin tissue. Even 16 months after cessation of UV exposure, accumulated damage was increased in some individuals by 32-fold (Berneburg et al., 2004). The causative role of oxidative stress for the increased frequency of mtDNA aberrations was demonstrated also in a variety of other experimental models and in human studies (Blatt et al., 2010). Also, a causative link between mtDNA mutations and aging phenotypes in mammals was evidenced in knock-out mice. Age-related phenotypes such as reduced subcutaneous fat, weight loss, alopecia, reduced fertility and heart enlargement were observed besides increased amount of point mutations and increased amount of deleted mtDNA (Trifunovic et al., 2004). Increased formation of oxidative stress due to endogenous or exogenous sources (e.g., UVR) causes mtDNA lesions in human cells and detrimental changes in mitochondrial respiration. Blatt et al. (2010) explain that due to a negative feedback loop, damaged mtDNA is a cause and a consequence of aging as well. Any lack of mitochondrial function impairs cellular ATP synthesis, thus reducing the "fuel supply" for repair mechanisms. It does further stimulate the formation of ROS as byproducts of an impaired mitochondrial respiration. More ROS are produced, more damage accumulates in neighboring mitochondrial complexes, membranes and mtDNA, progressive decline in the energy-generating capacity of the cell is observed, what further accelerates the aging process. Due to impaired ATP formation, the switch to anaerobic pathway such as glycolysis occurs, but also glycolysis additionally increases the formation of oxidative stress by the formation of reactive glycolytic intermediates, which favor the formation of advanced glycation end products (AGEs).

Thus even a single sunburn period could, in theory, start the above-described chain reaction leading to accelerated aging process in the skin cell.

UVR and ROS Formation

UVR can increase the ROS formation in the skin cells. Oxidative stress is believed to underlie changes associated with both photoaging and natural aging. According to Pattison and Davies (2006), UV radiation can mediate damage via two different mechanisms: (a) direct absorption of the incident light by the cellular components, resulting in excited state formation and subsequent chemical reaction, and (b) photosensitization mechanisms, where the light is absorbed by endogenous (or exogenous) sensitizers that are excited to their triplet states. The excited photosensitizers can induce cellular damage by two mechanisms: (a) electron transfer and hydrogen abstraction processes to yield free radicals (Type I); or (b) energy transfer with O2 to yield the reactive excited state, singlet oxygen (Type II) (Pattison and Davies, 2006).

The primary mechanism by which UV radiation initiates molecular responses in human skin is via photochemical generation of ROS mainly formation of superoxide anion ($O_2^{\cdot-}$), hydrogen peroxide (H_2O_2), hydroxyl radical ($OH^{\cdot}$), and singlet oxygen ($^1O_2$) (Hanson and Clegg, 2002). The main DNA product generated by (1)O_2 is 8-oxo-Gua; this is a common lesion in DNA and is formed by a range of other oxidants in addition to UV. UV light does not deposit sufficient energy in water molecules to fragment them, in contrast to X-rays and γ- rays. However, in the presence of H_2O_2 (formed during normal metabolism), UVB forms hydroxyl radical $OH^{\cdot}$.

$$H_2O_2 + UV \rightarrow OH^{\cdot} + OH^{\cdot}$$

In vitro, ex vivo and solution-phase studies have found that ROS, such as singlet oxygen ($^1O_2$), H_2O_2, superoxide ($O_2^{\cdot-}$) and nitric oxide, are generated after absorption of UV radiation by chromophores that are also found in keratinocytes (urocanic acid, riboflavin, reduced form of nicotinamide adenine dinucleotide–reduced nicotinamide adenine dinucleotide phosphate [NADH/ NADPH], tryptophan) (Hanson et al., 1997; Hanson et al., 1998).

UVA radiation produces cellular modifications that considerably overlap those induced by oxidative damage (Tyrrell, 1991). UVA radiation constitutes an oxidant stress that involves the generation of active species including singlet oxygen and hydroxyl radicals.

Hydrogen peroxide can be generated by UVA irradiation of tryptophan (McCormick et al., 1976), and superoxide can be produced by UVA irradiation of NADH and NADPH (Cunningham et al., 1985). The skin-damaging effects of UVA appear to result from type II, oxygen-mediated photodynamic reactions in which UVA or near-UV radiation in the presence of certain photosensitizing chromophores (e.g., riboflavin, porphyrins, nicotinamide adenine dinucleotide phosphate (NADPH), etc.) leads to the formation of reactive oxygen species ($^1O_2$, O_2^-, .OH) (Dalle Carbonare and Pathak, 1992).

Iron-free porphyrins also generate singlet oxygen upon exposure to UVA radiation. Also, other small molecules have the potential to generate active oxygen intermediates upon UVA exposure (Tyrell, 1992). For example, the photochemical degradation of tryptophan by wavelengths that include the more energetic portion of the UVA spectrum is able to generate hydrogen peroxide and N-formyl kynurenin (McCormick et al., 1976).

Although the level of hydrogen peroxide generated *in vivo* by such a pathway would appear to be in the low micromolar range, it could nevertheless be crucial to biological processes since iron complexes (such as citrate) that are present in the cytoplasm will react with hydrogen peroxide to generate the highly reactive hydroxyl radical in a superoxide driven Fenton reaction (Imlay et al., 1988). Since the reaction is driven by the continual reduction of ferric iron to the ferrous state by superoxide anions, a cellular source of superoxide anions is also required.

In this context, it should be noted that both hydrogen peroxide and hydroxyl radical are generated by UVA irradiation of NADH and NADPH (Czochralska et al., 1984; Cunningham et al., 1985). However, it is not at all clear whether this is really the key source of superoxide anions or whether the main source is as a consequence of normal cellular metabolism. Iron liberation and photodamaged skin contribute to oxidative damage and activation of transcription factors (Halliwell and Gutteridge, 2007).

Indeed, chronic exposure of hairless mice to low levels of UVB increases non-hem iron content of the skin; increases in iron with age and sun exposure are also observed in human skin. Topical application of certain iron ion chelators to skin of hairless mice appeared to delay the onset of UVB-induced damage (Halliwell and Gutteridge, 2007).

Naturally occurring iron complexes can react in vivo with hydrogen peroxide in the presence of superoxide anion in the Haber–Weiss reaction, which produces the potentially lethal hydroxyl radical.

Molecular Mechanisms by Which UVR Causes Photoaging

In previous paragraph, evidence was given that UV radiation generates reactive oxygen species and ROS can further oxidize cellular components. UV irradiation also directly or indirectly initiates and activates a complex cascade of biochemical reactions in human skin. Besides, the UV light-induced ROS interfere with signaling pathways. On a molecular level, UV radiation from the sun attacks keratinocytes and fibroblasts, resulting in the activation of cell surface receptors, which initiate signal transduction cascades. This, in turn, leads to a variety of molecular changes, which causes a breakdown of collagen in the extracellular matrix and a shutdown of new collagen synthesis (Fisher, 2005). UV-induced liberation of ROS in human skin is responsible for stimulation of numerous signal transduction pathways via activation of cell surface cytokine and growth factor receptors. UVA or UVB induce activation (sometimes via peroxides or singlet O_2 as signaling molecules) of a wide range of transcription factors in skin cells (Halliwell and Gutteridge, 2007). This can increase production of matrix metalloproteinases that can degrade collagen and other connective tissue components. For example, the UV light-induced ROS induce the transcription factor activator protein-1 (AP-1). AP-1 induces upregulation of matrix metalloproteinases (MMPs) like collagenase-1 (MMP-1), stromelysin-1 (MMP-3), and gelatinase A (MMP-2), which specifically degrade connective tissue such as collagen and elastin and indirectly inhibit the collagen synthesis in the skin (Rasche and Elstner, 2010). As indicated by their name, these zinc-dependent endopeptidases show proteoplytic activity in their ability to degrade matrix proteins such as collagen and elastin (Krutmann and Gilchrest, 2006). Destruction of collagen is a hallmark of photoaging. The major enzyme responsible for collagen 1 digestion is matrix metalloproteinase-1 (MMP-1) (Dong et al., 2008). Skin fibroblasts produce MMP-1 in response to UVB irradiation, and keratinocytes play a major role through an indirect paracrine mechanism involving the release of epidermal cytokine after UVB-irradiation (Fagot et al., 2002). MMPs are produced in response to UVB irradiation in vivo and are considered to be involved in the changes in connective tissue that occur in photoaging (Brinckmann et al., 1995). They are associated with a variety of normal and pathological conditions that involve degradation and remodeling of the matrix (Smutzer, 2002). UV rays and aging lead to excess proteolytic activity that disturbs the skin's three-dimensional integrity (Rocquet and Bonte, 2002).

These proteinases are important for breaking down the extracellular matrix during chronic wound repair, in which there is re-epithelialization by keratinocyte migration. Thus, MMPs are continuously involved in the remodeling of the skin after chronic aggression. Photodamage also results in the accumulation of abnormal elastin in the superficial dermis, and several MMPs have been implicated in this process (Rocquet and Bonte, 2002). ROS activate cytoplasmic signal transduction pathways in resident fibroblasts that are related to growth, differentiation, senescence, and connective tissue degradation (Scharffetter-Kochanek et al., 2000).

As well as causing permanent genetic changes involving protooncogenes and tumor suppressor genes, ROS activate cytoplasmic signal transduction pathways that are related to growth differentiation, senescence, transformation and tissue degradation (Scharffetter-Kochanek et al., 1997). The study of Kang et al. (2003) revealed that UVA/UVB irradiation of skin causes generation of H_2O_2 within 15 minutes. AP-1, which leads to increased collagen breakdown, becomes elevated and remains elevated within 24hours following UV irradiation (Fisher et al., 1996). Decreased procollagen synthesis within eight hours of UV irradiation was demonstrated (Quan et al., 2002). Consequently, increased collagen breakdown was demonstrated (Fisher et al., 1997). It is hypothesized that dermal breakdown is followed by repair that, like all wound repair, is imperfect. Imperfect repair yields a deficit in the structural integrity of the dermis, a solar scar. Dermal degradation followed by imperfect repair is repeated with each intermittent exposure to ultraviolet irradiation, leading to accumulation of solar scarring and ultimately visible photoaging (Fisher and Voorhees, 1998). While it may seem that the signs of photoaging appear overnight, they actually lie invisible beneath the surface of the skin for years. UV exposure of the skin causes oxidative stress, leading to inflammatory reactions, such as erythema, sunburn, and chronic reactions.

Most problematic chronic reactions include premature skin aging and skin cancer (Oresajo et al., 2010). The amount of photoaging that develops depends on: 1) a person's skin color and 2) their history of long-term or intense sun exposure. People with fair skin who have a history of sun exposure develop more signs of photoaging than those with dark skin.

It should be remembered that solar radiation includes not only UVR but also visible and infrared radiation. Visible light is thought to be unimportant in photoaging (Kligman, 1986), but studies have confirmed that infrared radiation can certainly damage the dermal matrix (Kligman and Kligman, 1984). It could be concluded that photoaging plays a significant role in skin-aging process.

Intermittent Exposure Hypothesis

Although the evidence from epidemiological studies indicates an association between melanoma and sunlight exposure, it does not appear that cumulative sun exposure explains the relationship, as it does for non-melanoma skin cancer (NMSC). Instead, an intermittent exposure hypothesis proposes that infrequent intense exposure of unacclimatized skin to sunlight is related to increasing melanoma incidence and is more important than chronic sun exposure. This hypothesis is supported by the observation that most studies have shown that an increased risk of melanoma is associated with a past history of severe sunburn in childhood and adolescence (Armstrong, 1988). This could be attributed also to the effect of hormesis, which explains action resulting from a response of an organism to low intensity of a stress (e.g., UVR exposure). Stress response then activates increased protection mechanisms and DNA repair systems.

According to World Health Organization - International Agency for Research on Cancer, in the Monographs on the evaluation of carcinogenic risks to humans' solar and ultraviolet, radiation is classified:

"Solar radiation is carcinogenic to humans (Group 1). There is sufficient evidence in humans for the carcinogenicity of solar radiation. Solar radiation causes cutaneous malignant melanoma and nonmelanocytic skin cancer. There is sufficient evidence for the carcinogenicity of solar radiation in experimental animals. There is sufficient evidence for the carcinogenicity of broad-spectrum ultraviolet radiation in experimental animals. There is sufficient evidence for the carcinogenicity of ultraviolet A radiation in experimental animals. There is sufficient evidence for the carcinogenicity of ultraviolet B radiation in experimental animals. There is *sufficient evidence* for the carcinogenicity of ultraviolet C radiation in experimental animals. Ultraviolet A radiation is probably carcinogenic to humans (Group 2A). Ultraviolet B radiation is probably carcinogenic to humans (Group 2A). Ultraviolet C radiation is probably carcinogenic to humans (Group 2A)."

Although DNA damage due to ROS is not a rare event since it is estimated that human cell sustains an average of 10^5 oxidative hits per day due to cellular oxidative metabolism (Fraga et al., 1991), DNA is functionally very stable, so that the incidence of cancer is much lower than one would expect, taking into account the high frequency of oxidative hits. Nevertheless, avoidance of excessive cumulative and sporadic sun exposure is important in reducing the risk of skin cancer and skin aging. Additionally, antioxidants might act by enhancing the DNA enzyme repair systems through a post-transcriptional gene regulation of transcription factors (Hirota et al., 1997). The repair capacity of human skin cells, therefore, directly relates to the probability of initiation of the carcinogenesis process and eventually tumor formation. Cellular antioxidant defense mechanisms are, therefore, crucial for the prevention or removal of the damage caused by the oxidizing component of UV radiation.

The most important strategy to reduce the risk of sun UV radiation damage is to avoid the sun exposure and to engage in the use of sunscreens. The next step is the use of exogenous antioxidants orally or by topical application and interventions in preventing oxidative stress and in enhanced DNA repair. Yarosh (2010) claims that properly conceived efforts to alleviate skin aging have the benefit of reducing rates of skin cancer since people are more motivated by improving their physical appearance than lowering their perceived risk of disease. Thus, the most successful anticancer efforts will arrive as treatments for skin aging.

Sunburn

Sunburn, or erythema, is an acute injury following excessive exposure to solar UVR. Sunburns are an acute inflammation reaction of the skin and tissue just beneath it that follows excessive exposure of the skin to UVR. The affected area becomes red, hot, tender, and swollen, and in severe cases, blisters may form. Low-dose or short exposure to UV irradiation is tolerated by the skin without noticeable or clinically relevant changes. Only after a certain threshold is reached does delayed and prolonged vasodilatation develop, allowing passage of lymphocytes and macrophages into the tissue and induction of an inflammatory response that is clinically visible as erythema. In epidemiology studies, sunburn is usually defined as burn with pain and/or blistering that lasts for two or more days. Cumulative sun exposure is the additive amount of sun exposure that one receives over a lifetime. A frequently used measure of UV irradiation–induced erythema is determination of the minimal erythema dose (MED). One MED is the minimal amount of energy required to induce a uniform, clearly demarcated redness 16 to 24 h after exposure to UV irradiation. Even a single minimum erythema dose (1 MED) can damage the dermal matrix. Cumulative sun exposure may reflect the additive effects of intermittent sun exposure, or chronic sun exposure, or both. The redness of the skin that results is due to an increased blood content of the skin by dilatation of the superficial blood vessels in the dermis, mainly the subpapillary venules (Diffey, 1991).

Skin color is an important factor in determining the ease with which the skin will sunburn. Whereas fair-skinned people require only about 15 to 30 min of midday summer sunshine to induce an erythemal reaction, people with moderately pigmented skin may require one to two hours exposure, and those with darkly pigmented skin will not normally sunburn. Other phenotype characteristics that may influence the susceptibility to sunburn are hair color, eye color and freckles (Azizi et al., 1988). Based on a personal history of response to 45 to 60 min of exposure to midday summer sun in early June (Fitzpatrick, 1988), individuals can be grouped into six sun-reactive skin types (Table 1). Types 1 and 2 are at high risk of skin cancer, particularly when exposed to intense sunlight.

There are anatomical differences in erythemal sensitivity (Diffey, 1991). The face, neck and trunk are two to four times more sensitive than the limbs. Vertical surfaces of an upright person receive about one-half of the ambient UVR, whereas horizontal surfaces, such as the epaulet region of the shoulder, receive up to 75%.

In addition to erythema and tanning, thickening (hyperplasia) of the epidermis is a significant component of a mild sunburn reaction (Diffey, 1991). A single moderate exposure to UVB can result in up to a three-fold thickening of the stratum corneum within one to three weeks, and multiple exposures every one to two days for up to seven weeks will thicken the stratum corneum by about three- to five-fold (Miescher, 1930). Skin thickness returns to normal about one to two months after ceasing irradiation.

Thickening of the skin (especially of the stratum corneum) after sun exposure can lead to a significant increase in protection against UVR by a factor of five or even higher. In Caucasians, skin thickening is probably more important than tanning in providing endogenous photoprotection, although in darkly pigmented races, skin pigmentation is the most important means of protection against solar UVR.

Table 1. Fitzpatrick's skin types

1.	Type I: Extremely fair skin, always burns, never tans.
2.	Type II: Fair skin, always burns, sometimes tans.
3.	Type III: Medium skin, sometimes burns, always tans.
4.	Type IV: Olive skin, rarely burns, always tans.
5.	Type V: Moderately pigmented brown skin, never burns, always tans.
6.	Type VI: Markedly pigmented black skin, never burns, always tans.

Fitzpatrick, 1988.

Tanning

Less intense or shorter duration exposure to UVR results in an increase in skin pigmentation that provides some protection against further UVR-induced damage. The increased skin pigmentation occurs in two phases, immediate pigment darkening and delayed tanning. Intermediate pigment darkening occurs during exposure to UVR and results from oxidation and redistribution of existing melanin. This reaction may fade rapidly or persist for several days. Delayed tanning results from increased synthesis of epidermal melanin and requires 24 to 72 hours to become visible.

Melanocytes are specialized dendritic cells interspersed among basal keratinocytes and serve the primary function of producing melanin in intracellular organelles melanosomes that are then distributed to surrounding keratinocytes (Hakozaki et al., 2010).

Following solar UVR exposure, there is an increase in the number of functioning melanocytes, and activity of the enzyme tyrosinase is enhanced (Fitzpatrick et al., 1983). This leads to the formation of new melanin and hence an increase in the number of melanin granules throughout the epidermis. Melanins are the major UV-absorbing chromophores in skin, exhibiting an extremely broad spectrum of absorption over the UVB, UVA and visible ranges. Melanins are complex polymeric proteins that are produced by melanocytes and transferred to keratinocytes. It should be stressed that melanin cannot offer 100% protection to our skin against harmful effects of solar radiation.

Gilchrest et al., (1999) explained how UVR stimulates melanogenesis in the skin. A direct effect of UV photons on DNA results in upregulation of the gene for tyrosinase, the rate-limiting enzyme in melanin synthesis, as well as an increase in cell-surface expression of receptors for at least one of the several known keratinocyte-derived melanogenic factors, MSH. Direct effects of UV on melanocyte membranes, releasing arachidonic acid, may also play a role in the tanning response. The tanning response also relies heavily on UV-stimulated increased production and release of numerous keratinocyte-derived factors including bFGF, NGF, endothelin-1 and the POMC-derived peptides MSH, ACTH, beta-LPH and beta-endorphin (Gilchrest et al., 1999). Ultraviolet-induced melanogenesis may be one part of a eukaryotic SOS response to damaging UVR that has evolved over time to provide a protective tan in skin at risk of further injury from sun exposure (Gilchrest et al., 2006).

The distribution and size of melanin particles also plays an important role in protecting epidermal cells.

Melanin particles have a distribution within the stratum corneum and epidermal cells depending upon skin type. In dark skin types (5 and 6), these particles are positioned within cells to provide optimum optical protection for the cell nuclei and in adequate size in the stratum corneum (Kollias et al., 1991).

According to Euro melanoma (http://www.euromelanoma.org/uk/home) "fake tan (auto bronzing) is by far safer than a suntan. A fake tan produces a natural looking tan through a chemical reaction in the skin. It, therefore, gives the desired cosmetic result in someone who is very keen to have a tan, without him or her having to sunbathe. Application of fake tan needs to be carried out every one to two weeks in order to maintain the tan since the top layers of the skin are constantly being renewed. However, the actual tanning produces by 'fake tan' is not at all protective against UV-radiation and consumers should be aware of this important factor."

Beneficial Effects of UV

Until now, just harmful effects of UVR on different cellular molecules of the skin were discussed. However, small amounts of UV are beneficial for people and essential in the production of vitamin D. UVR is also used to treat several diseases, including rickets, psoriasis, atopic dermatitis and jaundice.

This takes place under medical supervision, and the benefits of treatment versus the risks of UVR exposure are a matter of clinical judgment.

According to Diffey (1991), the only thoroughly established beneficial effect of solar UVR on the skin is the synthesis of vitamin D3. Solar radiation in the UVB waveband photochemically converts 7-dehydrocholesterol in the epidermis to previtamin D3. This previtamin immediately isomerizes to vitamin D3 in a reaction controlled by skin temperature and which takes two to three days to reach completion. Previtamin D3 is photolabile, and excessive exposure to sunlight causes its photolysis to biologically inert photoproducts, lumisterol and tachysterol. In fact, production of previtamin D3 is limited to no more than 5% to 15% of the total 7-dehydrocholesterol content in the skin, no matter how long a person is exposed to sunlight. Once vitamin D3 is made in the skin, it enters the blood for transport to the liver to be metabolized to 25-hydroxyvitamin D (Webb and Holick, 1988). If vitamin D3 does not enter the circulation before sun exposure the following day, it can be rapidly degraded in the skin by sunlight to suprasterol 1, suprasterol 2 and 5,6-transvitamin D3—products which are believed to be biologically inert (Webb et al., 1989). Thus sunlight, through its photochemical activity, is able to regulate the production of both previtamin D3 and vitamin D3 in the skin.

Only short exposures to sunlight are required to synthesize vitamin D3 in the skin; from spring until autumn, 15 min exposure to the hands, arms and face between 9 am and 4 pm is adequate to provide our vitamin D3 requirement (Diffey, 1991). There is still the ongoing debate between the adequate sun exposure to create sufficient vitamin D and the risk arising from skin cancer from moderate increased sun exposure (Grant and Holick, 2005). It has been roughly estimated that a sufficiency of vitamin D synthesis in skin can be provided by exposure of 40% of the body to 25% of the UVB minimal erythemal dose (Grant and Holick, 2005). An untanned person with fair skin may receive mild sunburn in as little as 25 minutes at noon (depending on the time of year and the latitude) but would have to lie in the sun for at least two hours to receive the same dose after 3 pm (International program on chemical safety, Environmental Health Criteria 160). Alternative to sun exposure vitamin D production is the intake of synthetic vitamin D with supplements. But on the other hand, supplements of vitamin D can cause side effects from excess vitamin D, while sun-exposure cannot lead to excessive vitamin D amounts. However, it can lead to excessive skin damage formation.

Harmful Effects of Sunscreens

Sunscreens contain chemical organic filters, which mainly absorb UVB or UVA radiation, and physical filters (TiO_2 and ZnO), which block the UVB/ UVA part of the solar radiation by reflection and scattering (Diaz-Cruz et al., 2008, Food and Drug Association (FDA) documents http://www.cfsan.fda.gov/~lrd/fr990521.html; Salinaro et al., 1999; Serpone et al. 1999). Such products must be photostable in order to spend energy effectively mainly through photophysical processes and minimize photochemical ones which involve singlet oxygen or other ROS or intermediates. They should not penetrate deeper into the skin, and also may not be transported into the cells. Alternatively, they should prevent UVB and UVA rays from penetrating into the cell nucleus and damage DNA. Products for skin protection containing UV filters are in the most countries declared as "cosmetic products" (Klein, 1992; Lowe, 1990). Sunscreens should always contain these compounds in combination, because there is no single UV filter, which at present would provide a sufficiently high SPF according to the legislation (Serpone, 2007). In the formulation of these compounds the FDA recommends a minimum of 2 mg lotion/cm^2 skin present at the ratio of mg/g to be applied to the skin or hair in large quantities.

Organic UV filters, also regarded as chemical UV filters which are responsible for absorption of the solar UV radiation, comprise different classes of compounds such as UVA (benzophenones, anthranilates and dibenzoylmethane) and UVB filters (PABA derivatives, salicylates, cinnamates and camphor derivatives) that block radiation in UVA or UVB (Díaz-Cruz et al., 2008). Dibenzoylmethane derivatives are widely used as organic UVA filters, among them the most common is butyl methoxydibenzoylmethane (avobenzone). With the increase of industrial production of nano and microparticles, these products have found applications in sunscreens as well. But still, questions arise about its effects on humans and animals (Yeats and Mauderly, 2001). The European Commission as early as 2006 estimated that at the market level there was about 5 % of cosmetic products with nanoparticles.

Previous research has shown that some organic UV filters decompose when exposed to light. In the presence of chlorine and chlorinated medium, like water pools or seawater, direct photolytic reactions or chlorination of the aromatic ring or the side chain, may also occur. Chlorination is the most commonly used chemical process for disinfecting swimming pools and drinking water. The formation of halogenated by-products in chlorinated waters is inevitable, especially when they are substituted with phenolic compounds and/or amino groups (Bedner and MacCrehan, 2006; Lebedev, 2007; Duirk et al., 2013; Zhuang et al., 2013; Grbović et al., 2013). Direct release of applied sunscreens, as consequence of recreational activities in swimming pools and sunbathing areas, and indirect discharge through domestic wastewater (e.g. during showering, clothes washing) represent the main pathways for organic UV filters to enter the aquatic environment (Díaz-Cruz et al., 2008). Information on detected environmental concentration of organic UV filters in freshwater (i.e. rivers, lakes) and marine ecosystems are summarized in numerous recent publications and are mainly in ng L^{-1} range, reaching often levels up to 4-10 and even 24 μg L^{-1} for specific UV filtering compound (Fent et al., 2010; Kaiser et al., 2012; Santos et al., 2012; Duirk et al., 2013; Xiao et al., 2013). On contrary, data regarding the presence of organic UV filters in swimming pool water are rather limited. Vidal et al. (2010), Cuderman and Heath (2007) reported values for isoamyl *p*-methoxycinnamate, benzophenone 3 and 4-methylbenzylidene camphor of 700 ng L^{-1}, 400 ng $L^{-1}$and 330 ng L^{-1}, respectively. Higher content of benzophenone 3, Eusolex 6300 (4-methylbenzylidene camphor) and Eusolex 2292 (octyl-methoxycinnamate) in swimming pool waters was determined by Lambropoulou et al. (2002) and Giokas et al. (2004) with concentrations from 2 to 10 μg L^{-1}. Zhuang et al. (2013) reported about presence of benzophenone 3 in swimming pool waters in concentration of 0.3 and 1.7 μg L^{-1} and about the presence of 3,5-dichloro benzophenone 3 in the concentration of 6.6 μg/L.

In the last decade there has been extensive discussion regarding the quality of swimming pool waters. Opinions from different experts representing different sectors are quite opposite. On one side some of them claim, that the quality of swimming waters is constantly improving, but on the other side some others warn about the presence of various compounds pool waters contain, from cosmetics to pharmaceuticals (especially residues and metabolites) and other contaminants, and about the possible effects of water consumption (also as inhalation of aerosols), which means increased health risk including an increased incidence of cancer. A special problem represents chlorinated products formed under conditions of disinfectant (e.g. swimming pool water) since these compounds are new ones and no literature on effects of chlorinated compounds are available.

Endocrine disrupting activity of small sized organic and nano sized inorganic UV filters have been reported. In the literature we found very little data pertaining to estrogen, anti-estrogen, androgen and antiandrogen effects of parent organic UV filters (Fent et al., 2010; Zucchi et al., 2011). Due to the high lipophilicity these compounds tend to accumulate or bioaccumulate in the sediments as well as in the food chain. Some UV filters have been detected in human breast milk (Schlumpf et al., 2010) and urine (Leon et al. 2010). Just a few research studies are available on the transformation characteristics and potential health risks of benzophenone-type UV filters during chlorination disinfection process (Liu et al, 2014). Data on the effects of these compounds on the skin structure in the scientific literature is not available. Certain organic UV filters (PABA derivatives, cinnamates, benzophenones, and octocrylene) have been described to cause photoallergy.

Recent studies have shown on allergic and photoallergic reactions of skin exposed to UV filters. One of epidemiological studies has shown that avobenzone and some other UV filters cause photoallergic dermatitis (Gaspar et al., 2013). Many sunscreens also contain antioxidants (vitamin A, C), which act as photoprotection and maintain or restore a healthy skin barrier, but at the same time can also form toxic photoproducts and allergens.

The cosmetic industry is increasingly using new organic compounds and nanoparticles in products for protection against UV irradiation. For the safety of cosmetic products the responsibility is on the companies themselves.

The testings' are carried out by them or by authorized laboratories. Currently, existing legislation does not provide clear guidance particularly with regard to the use of nanoparticles in cosmetics. Cosmetic companies, therefore, use tests that apply to other substances, or avoid the use of nanoparticles or do not advertise their addition. A potential risk of vitamin D deficiency by sunscreen use has become a major subject of public health debate (Bens, 2013) since 1,25 (OH) 2D stimulates the differentiation and inhibits the proliferation of cells, although conclusive evidence of vitamin D in cancer prevention it is not yet definite (Shui and Giovannucci, 2013). Already SPF of 8 inhibits or even significantly prevents vitamin D production in the skin (Holick, 2004; Sayre and Dowdy, 2007). Additionally, reducing the exposure of the skin UVB radiation could suppress the skin's production of melanin (Autier et al., 1995; Meredith and Riesz, 2004).

Treatment of Photoaging

The primary treatment of photoaging is photoprotection, however the additional treatment could be achieved with the use of antioxidants as well with some novel compounds such as polyphenols. Exogenous antioxidants like vitamin C, E and many others cannot be synthesized by the human body and must be taken up by the diet. Natural antioxidants are generally considered to be beneficial fruit and vegetable components. It seems that skin's antioxidative defense is also influenced by nutritive factors.

Besides vitamins A, C, and E, η-3 fatty acids certain non-vitamin plant derived ingredients might have beneficial effect on skin aging, skin sun protection, or skin cancer. The laboratory studies conducted in animal models suggest that many plant compounds have the ability to protect the skin from the adverse effects of UVR (Pandel et al., 2013; Bruce, 2008).

Many studies have found that vitamin C can increase collagen production, protect against damage from UVA and UVB rays, correct pigmentation problems, and improve inflammatory skin conditions (Poljšak, 2011). Ascorbic acid was a photoprotectant when applied to mice and pig skin before exposure to ultraviolet (UV) radiation (Elmor, 2005).

The application of retinoids might not only clinically and biochemically repair photoaged skin, but their use might also prevent photoaging (Serri and Iorizzo, 2008). Supplements or a carotenoid-rich diet decreased sensitivity against UV-induced erythema.

Supplementation with carotenoids contributes to basal protection of the skin but is not sufficient to obtain complete protection against severe UV irradiation. Beta-carotene acts not only as an antioxidant but also has unexpected prooxidant properties (Biesalski and Obermueller-Jevic, 2001). A number of experimental studies indicate protective effects of beta-carotene against acute and chronic manifestations of skin photodamage. For this reason, further studies with focus on *in vivo* β-carotene-induced prooxidative properties and its relevance on human health are needed (Pandel et al., 2013).

It was reported that CoQ10 strongly inhibits oxidative stress in the skin induced by UVB via increasing SOD2 and GPx (Kim et al., 2007). A topical application of CoQ10 has the beneficial effect of preventing photoaging (Hoppe et al., 1999) as well as protecting skin against oxidative stress-induced cell death and enhances the synthesis of basement membrane components in dermal and epidermal cells (Muta-Takada et al., 2009). Furthermore, CoQ10 appears to have also a cutaneous healing effect *in vivo* (Choi et al., 2009).

Green tea polyphenols have received attention as protective agents against UV-induced skin damage. Analysis of published studies demonstrates that green tea polyphenols have anti-inflammatory and anticarcinogenic as well as antiaging properties (Katiyar et al., 2000).

Topical treatment or oral consumption of green tea polyphenols (GTP) inhibits chemical carcinogen or UV radiation-induced skin carcinogenesis in different laboratory animal models (Katiyar, 2003). Green tea itself or caffeine in amounts equivalent to three of five cups of coffee per day to UVB-exposed mice increased levels of p53, slowed cell cycling, and increased apoptotic sun burn cells in the epidermis (Lu et al., 2008).

Dietary flavanols from cocoa contribute to endogenous photoprotection, improve dermal blood circulation, and affect cosmetically relevant skin surface and hydration variables (Heinrich et al., 2006), but it was reported too that the Cocoa photoprotection against UV-induced erythema (Heinrich et al., 2006).

Conclusion

After skin damage due to sun exposure, repair mechanisms are not 100% effective. The damaged components are not always completely repaired.

The problem arises in the cases of intensive acute sun exposure or in the cases of chronic sun exposure over longer decades, which manifests as skin photoaging.

To what extent each mechanism—mtDNA mutagenesis, protein oxidation, down regulation of collagen synthesis and increased expression of matrix metalloproteinases—contributes to premature skin aging is still not answered.

As photoaging via sunlight or artificial UV-exposure is the major impacting factor for skin appearance, new defense strategies have been suggested, including the appropriate UVA+UVB sunscreen choice in addition to an antioxidant-rich diet, the induction of photoprotective melanogenesis by means of thymidine-dinucleotide formulations, the use of phytoestrogens and metal chelating agents to inhibit the collagenase activation, the avoidance of refined hyperglycemic carbohydrates so as to slow down the glycation/ oxidation of proteins, the use of aminoguanidine and carnosine formulations to inhibit the collagen cross-linking as well as retinoic acid to stimulate the DNA repair mechanisms and collagen synthesis (Ionescu, 2005).

UV protection includes not only reduction of sun exposure but also use of sun protective filters, UV protective clothes, DNA repair enzymes, and antioxidant supplementation.

References

Anderson, L. (2006). Looking Good, the Australian guide to skin care, cosmetic medicine and cosmetic surgery. *AMPCo*. Sydney. ISBN 0-85557-044-X.

Armstrong, B. K. (1988). Epidemiology of malignant melanoma: intermittent or total accumulated exposure to the sun? *J. Dermato., Surg. Oncol.*, 14, 835-49.

Azizi, E., Lusky, A., Kushelevsky, A. P., Schewach-Millet, M. (1988). Skin type, hair color, and freckles are predictors of decreased minimal erythema ultraviolet radiation dose. *J. Am. Acad. Dermatol.*, 19, 32-8.

Bedner, M., MacCrehan, W. A. (2006). *Environ Sci. Technol.*, 40, 516-22.

Berneburg, M., Plettenberg, H., Medve-König, K., Pfahlberg, A., Gers-Barlag, H., Gefeller, O., Krutmann, J. (2004). Induction of the photoaging-associated mitochondrial common deletion in vivo in normal human skin. *J. Invest. Dermatol.*, 122, 1277-84.

Bernstein, E. F., Chen, Y. Q., Kopp, J. B., Fisher, L., Brown, D. B., Hahn, P. J., et al. (1996). Long-term sun exposure alters the collagen of the papillary dermis. Comparison of sun-protected and photoaged skin by northern analysis, immunohistochemical staining, and confocal laser scanning microscopy. *Journal of the American Academy of Dermatology*, 34(2 Pt 1), 209-218.

Bhattacharyya, T. K. (2010). Skin Aging in animal models: histological perspective. In: *Textbook of skin aging*. Farage, M. A., Miller, K. W., Maibach (Eds.). Berlin: Springer-Verlag.

Biesalski, H. K., Obermueller-Jevic, U. C. (2001). UV light, beta-carotene and human skin—beneficial and potentially harmful effects. *Arch. Biochem. Biophys.*, 389(1), 1-6.

Blatt, T., Wenck, H., Wittern, K. P. (2010). Alterations of energy metabolism in cutaneous aging. In: *Textbook of aging skin*. Farage, M. A., Miller, K. W., Maibach, H. I. (Eds.). Berlin: Springer-Verlag.

Blumthaler, M., Ambach, W., (1985). Neure Messugen der Albedo verschiedener Oberflaechen fur erythemwirksame Strahlung. *Annal. Meteorol.*, 22, 114-5.

Bodkin, N. L., Ortmeyer, H. K., Hansen, B. C. (1995). Long-term dietary restriction in older-aged rhesus monkeys: effects on insulin resistance. *J. Gerontol. A Biol. Sci. Med. Sci.*, 50(3), B142-7.

Bolognia, J. L. (1993). Dermatologic and cosmetic concerns the older women. *Clinic in Geriatric Medicine*, 9, 209-29.

Brinckmann, J., Acil, Y., Wolff, H. H., Müller, P. K. (1995). Collagen synthesis in (sun)-aged human skin and in fibroblasts derived from sun-exposed and sun-protected body sites. *J. Photochem. Photobiol.*, 27, 33-8.

Bruce, S. (2008). Cosmeceuticals for the attenuation of extrinsic and intrinsic dermal aging. *J. Drugs Dermatol.*, 7(2 Suppl.): s17-22.

Bruls, W. A. G., Slaper, H., van der Leun, J. C., Berrins, L. (1984). Transmission of human epidermis and stratum corneum as a function of thickness in the ultraviolet and visible wavelengths *Photochem. Photobiol.*, 40, 485-94.

Cadet, J., Berger, M., Douki, T., Morin, B., Raoul, S., Ravanat, J. L., Spinelli, S. (1997). Effects of UV and visible radiation on DNA-final base damage. *Biol. Chem.*, 378(11), 1275-86.

Choi, B. S., Song, H. S., Kim, H. R., Park, T. W., Kim, T. D., Cho, B. J., Kim, C. J., Sim, S. S. (2009). Effect of coenzyme Q10 on cutaneous healing in skin-incised mice. *Arch. Pharm. Res.*, 32(6), 907-13.

Cuderman, P. E. (2007). *Heath, Analytical and Bioanalytical Chemistry*, 387, 1343-50.

Cunningham, M. L., Krinsky, N. I., Giovanazzi, S. M., Peak, M. J. (1985). Superoxide anion is generated from cellular metabolites by solar radiation and its components. *Free Radic. Biol. Med.*, 5, 381-5.

Cutchis, P. (1980). *A formula for comparing annual damaging ultraviolet (DUV) radiation doses at tropical and mid-latitude sites Final Report FAA-EE 80-21*. Washington, DC: US Department of Transportation, Federal Aviation Administration Office of Environment and Energy.

Czochralska, B., Bartosz, W., Shugar, D. (1984). Oxidation of excited-state NADH and NAD dimer in aqueous medium - involvement of O2- as a mediator in the presence of oxygen. *Biochim. Biophys. Acta*, 801, 403-9.

Dalle Carbonare, M., Pathak, M. A. (1992). Skin photosensitizing agents and the role of reactive oxygen species in photoaging. *J. Photochem. Photobiol. B.*, 14(1-2), 105-24.

Diaz-Cruz, M. S., Llorca, M., Barcelo, D. (2008). *TrAC-Trends in Analytical Chemistry*, 27 (10), 873–87.

Diffey, B. L., Farr, P. M. (1988). The action spectrum in drug-induced photosensitivity. *Photochem. Photobiol.*, 47, 49-54.

Diffey, B. L. (1991). Solar ultraviolet radiation effects on biological systems. *Review in Physics in Medicine and Biology*, 36(3), 299-328.

Doda, D. D., Green, A. E. (1981). Surface reflection measurements in the UV from an airborne platform. Part 2 *Appl. Opt.*, 20, 636-42.

Duirk, S. E., Bridenstine, D. R., Leslie, D. C. (2013). *Water Res.*, 47, 579-87.

Elmore, A. R. (2005). Final report of the safety assessment of L-Ascorbic Acid, Calcium Ascorbate, Magnesium Ascorbate, Magnesium Ascorbyl Phosphate, Sodium Ascorbate, and Sodium Ascorbyl Phosphate as used in cosmetics. *Int. J. Toxicol.*, 24(2), 51-111.

Ernster, V. L., Grady, D., Miike, R., Black, D., Selby, J., Kerlikowske, K. (1995) Facial wrinkling in men and women, by smoking status. *Am. J. Public Health*, 85, 78-82.

Fagot, D., Asselineau, D., Bernerd, F. (2002). Direct role of human dermal fibroblasts and indirect participation of epidermal keratinocytes in MMP-1 production after UVB irradiation. *Arch. Dermatol. Res.*, 293, 576-83.

Farage, M. A., Miller, K. W., Elsner, P., Maibach, H. I. (2008). Intrinsic and extrinsic factors in skin ageing: a review. *Int. J. Cosmet. Sci.*, 30(2), 87-95.

Farman, J. C., Gardiner, B. G., Shaklin, J. D. (1985). Large losses of total ozone in Antarctica reveal seasonal ClOx/NOx interaction. *Nature*, 315, 207-10.

Fent, K., Kunz, P. Y., Zenker, A., Rapp, M. (2010). *Mar. Environ. Res.* 69(S4-6).

Fisher, G. J., Kang, S., Varani, J., Bata-Csorgo, Z., Wan, Y., Datta, S., Voorhees, J. J. (2002). Mechanisms of photoaging and chronological skin aging. *Arch. Dermatol.*, 138(11), 1462-70.

Fisher, G. J. (2005). "The Pathophysiology of Photoaging of the Skin." *Cutis*, 75(2S), 5-9.

Fisher, M. S., Kripke, M. L. (1977). Systemic Alteration Induced in Mice by Ultraviolet Light Irradiation and its Relationship to Ultraviolet Carcinogenesis," *Proc. Natl. Acad. Sci. US* 74, 1688-92.

Fisher, G. J., Voorhees, J. J. (1998). Molecular mechanisms of photoaging and its prevention by retinoic acid: ultraviolet irradiation induces MAP kinase signal transduction cascades that induce Ap-1-regulated matrix metalloproteinases that degrade human skin in vivo. *J. Investig. Dermatol. Symp. Proc.*, 3(1), 61-8.

Fisher, G. J., Wang, Z. Q., Datta, S. C., Talwar, H. S., Wang, Z. Q., Varani, J., Kang, S., Voorhees, J. J. (1996). Molecular basis of sun-induced premature skin ageing and retinoid antagonism. *Nature*, 379, 335-9.

Fitzpatrick, T. B. (1988). The validity and practicality of sun-reactive skin types I through VI. *Arch. Dermatol.*, 124(6), 869-71.

Fraga, C. G., Motchnik, P. A., Shigenaga, M. K., Helbock, H. J., Jacob, R. A., Ames, B. N. (1991). Ascorbic acid protects against endogenous oxidative DNA damage in human sperm. *Proc. Natl. Acad. Sci. US* 88(24), 11003-6.

Frederick, J. E. (1990). Trends in atmospheric ozone and ultraviolet radiation: mechanisms and observations for the Northern Hemisphere. *Photochem. Photobiol.*, 51, 757-63.

Gaspar, L. R., Tharmann, J., Campos, P. M. B. G. M., Liebsch, M. (2013). *Toxicology in vitro*, 27(1), 418-25.

Gilchrest, B. A. (1998). Overview of skin aging. *J. Cut. Aging and Cos. Derm.*, 1(1), 1-2.

Gilchrest, B. A. (1990). Skin aging and photoaging. *Dermatology Nursing*, 2 (2), 79-82.

Gilchrest, P., Schieke, S., Morita, A. (2006). Premature skin aging by infrared radiation, tobacco smoke and ozone. In: *Skin aging*. (Gilchrest, B., Krutmann, J. (Eds.). Berlin: Springer-Verlag.

Giokas, D. L., Sakkas, V. A., Albanis, T. A. (2004). *J. Chromatogr. A,* 1026, 289-93.

Goukassian, D., Gad, F., Yaar, M., Eller, M. S., Nehal, U. S., Gilchrest, B. A. (2000). Mechanisms and implications of age-associated decrease in DNA repair capacity. *FASEB J.*, 14, 1325-34.

Grant, W. B., Garland, C. F., Holick, M. F. (2005). Comparisons of estimated economic burdens due to insufficient solar ultraviolet irradiance and vitamin D and excess solar UV irradiance for the United States. *Photochem. Photobiol.*, 81, 1276-86.

Grbović, G., Trebše, P., Dolenc, D., Lebedev, A. T., Sarakha, M. (2013). LC/ MS study of the UV filter hexyl 2-[4-(diethylamino)-2-hydroxybenzoyl]-benzoate (DHHB) aquatic chlorination with sodium hypochlorite. *J. Mass Spectrom.*, 48(11), 1232-40.

Grether-Beck, S., Wlaschek, M., Krutmann, J., Scharffetter-Kochanek, K. (2005). Photodamage and photoaging-prevention and treatment. *J. Dtsch. Dermatol. Ges.*, 3(2), S19-25.

Hakozaki, T., Swanson, C. L., Bissett, D. L. (2010). Hyperpigmentation in aging skin. In: Farage, M. A., Miller, K. W., Maibach, H. I. *Textbook of aging skin.* Berlin: Springer-Verlag.

Halliwell, B., Gutteridge, J. (1999). *Free radicals in biology and medicine* (3rd edn.). Oxford: Clarendon Press, 1999.

Halliwell, B., Gutteridge, J. (2007). *Free radicals in biology and medicine* (4th edn). Oxford: University Press, 2007.

Hanson, K. M., Simon, J. D. (1998) Epidermal trans-urocanic acid and the UVA-induced photoaging of the skin. *Proc. Natl. Acad. Sci. US*, 95, 10576–10578.

Hanson, K. M., Li, B., Simon, J. D. (1997). A spectroscopic study of the epidermal chromophore trans-urocanic acid. *J. Am. Chem. Soc.*, 119, 2715–2721.

Heinrich, U., Gärtner, C., Wiebusch, M., Eichler, O., Sies, H., Tronnier, H., Stahl, W. (2003). Supplementation with beta-carotene or a similar amount of mixed carotenoids protects humans from UV-induced erythema. *J. Nutr.*, 133(1), 98-101.

Heinrich, U., Neukam, K., Tronnier, H., Sies, H., Stahl, W. (2006). Long-term ingestion of high flavanol cocoa provides photoprotection against UV-induced erythema and improves skin condition in women. *J. Nutr.*, 136(6), 1565-9.

Helfrich, Y. R., Sachs, D. L., Voorhees, J. J. (2008). Overview of skin aging and photoaging. *Dermatol. Nurs.*, 20(3), 177-83.

Hirota, K., Matsui, M., Iwata, S., Nishiyama, A., Mori, K., Yodoi, J. (1997). AP-1 transcriptional activity is regulated by a direct association between thioredoxin and Ref-1. *Proc. Natl. Acad. Sci. US*, 94(8), 3633-8.

Hoppe, U., Bergemann, J., Diembeck, W., Ennen, J., Gohla, S., Harris, I., Jacob, J., Kielholz, J., Mei, W., Pollet, D., Schachtschabel, D., Sauermann, G., Schreiner, V., Stäb, F., Steckel, F. (1999). Coenzyme Q10, a cutaneous antioxidant and energizer. *Biofactors.*, 9(2-4), 371-8.

Hutchinson, F. (1985). Chemical changes induced in DNA by ionizing radiation. *Prog. Nucleic Acid Res. Mol. Biol.*, 32, 115-154.

Imlay, J. A., Chir, S. M., Linn, S. (1988). Toxic DNA damage by hydrogen peroxide through the Fenton reaction in vivo and in vitro. *Science*, 240, 640-642.

International programme on chemical safety, Environmental Health Criteria, 160, Ultraviolet radiation (EHC *160*, 1994, 2nd edition), accessed: www.inchem.org/documents/ehc/ ehc/ehc160.htm.

IARC, 2010???

Jenkins, G., Wainwright, L. J., Green, M. (2009). Healthy aging: The skin. In: *Healthy aging. The role of nutrition and lifestyle.* (Ed: Stanner, S., Thompson, R., Buttriss, J. L.). Wiley-Backwell.

Kaiser, D., Sieratowicz, A., Zielke, H., Oetken, M., Hollert, H., Oehlmann, J. (2012). *Environmental Pollution*, 163, 84-90.

Kang, S., Chung, J. H., Lee, J. H., Fisher, G. J., Wan, Y. S., Duell, E. A., Voorhees, J. J. (2003). Topical N-acetyl cysteine and genistein prevent ultraviolet-light-induced signaling that leads to photoaging in human skin in vivo. *J. Invest. Dermatol.*, 120(5), 835-41.

Katiyar, S. K., Ahmad, N., Mukhtar, H. (2000). Green tea and skin. *Arch. Dermatol.*, 136, 989–94.

Katiyar, S. K. (2003). Skin photoprotection by green tea: antioxidant and immunomodulatory effects. *Curr. Drug Targets Immune Endocr. Metabol. Disord.*, 3(3), 234-42.

Kim, D. W., Hwang, I. K., Kim, D. W., Yoo, K. Y., Won, C. K., Moon, W. K., Won, M. H. (2007). Coenzyme Q_{10} effects on manganese superoxide dismutase and glutathione peroxidase in the hairless mouse skin induced by ultraviolet B irradiation. *Biofactors.*, 30(3), 139-47.

Klein, K. (1992). *Cosmet. Toiletries*, 107, 45.

Kligman, L. H., Kligman, A. M. (1984). Reflections on heat. *Br. J. Dermatol.*, 110, 369-75.

Kligman, L. H., Kligman, A. M. (1986). The nature of photoaging: Its prevention and repair. *Photodermatology*, 3, 215-27.

Kollias, N., Sayre, R. M., Zeise, L., Chedekel, M. R. (1991). Photoprotection by melanin. *J. Photochem. Photobiol. B.*, 9(2), 135-60.

Krutman, J., Gilchrest, B. A. (2006). Photoaging of skin. In: *Skin aging*, Gilchrest, B., Krutmann, J. (Eds.). Berlin: Springer-Verlag.

Kushelevsky, A. P., Slifkin, M. A. (1975). Ultraviolet measurements at the Dead Sea and Beersheba Isr. *J. Med. Sci.*, 11, 488-90.

Lambropoulou, D. A., Giokas, D. L., Sakkas, V. A., Albanis, T. A., Karayannis, M. I. (2002). *J. Chromatogr.* A, 967, 243-53.

Laurent-Applegate, L. E., Schwarzkopf, S. (2001). Photooxidative Stress in Skin and Regulation of Gene Expression. In: *Environmental stressors in health and disease* (Ed. Fuchs, J., Packer, L.). Marcel Dekker.

Lavker, R. M. (1995). Cutaneous aging: Chronologic versus photoaging. In: B. Gilchrest (Ed.), *Photodamage* (Vol. 1, pp. 123-135). Cambridge: Blackwell.

Lebedev, A. T. (2007). *Eur. J. Mass Spectrum.*, 13(N1), 51-6.

León, Z., Chisvert, A., Tarazona, I., Salvador, A. (2010). *Analytical and Bioanalytical Chemistry*, 398(2), 831-43.

Liu, Y., Hong, L., Kempf, V. R., Wakamatsu, K., Ito, S., Simon, J. D. (2004). Ion-exchange and adsorption of Fe(III) by Sepia melanin. *Pigment Cell Research*, 17(3), 262–9.

Lowe, N. J., Shaath, N. A. (1990). *Cosmet. Sci. Technol. Ser.*, 10, 370.

Lu, Y. P., Lou, Y. R., Peng, Q. Y., Xie, J. G., Nghiem, P., Conney, A. H. (2008). Effect of caffeine on the ATR/Chk1 pathway in the epidermis of UVB-irradiated mice. *Cancer Res.*, 68(7), 2523-9.

Lu, Y. P., Lou, Y. R., Peng, Q. Y., Xie, J. G., Nghiem, P., Conney, A. H. (2008). Effect of caffeine on the ATR/Chk1 pathway in the epidermis of UVB-irradiated mice. *Cancer Res.*, 68, 2523-9.

McCormick, J. P., Fisher, J. R., Pachlatko, J. P., Eisenstark, A. (1976). Characterization of a cell lethal product from the photooxidation of tryptophan: hydrogen peroxide. *Science*, 198, 468-9.

Miescher, G. (1930). Das problem des Lichtschutzes und der Lichtgewohnung. *Strahlentherapie*, 35, 403-43.

Molina, M. J., Rowland, F. S. (1974). Stratospheric sink for chlorofluoromethanes: chlorine atom-catalyzed destruction of ozone. *Nature*, 249, 810-2.

Muta-Takada, K., Terada, T., Yamanishi, H., Ashida, Y., Inomata, S., Nishiyama, T., Amano, S. (2009). Coenzyme Q10 protects against oxidative stress-induced cell death and enhances the synthesis of basement membrane components in dermal and epidermal cells. *Biofactors.*, 35(5), 435-41.

Oresajo, C., Yatskayer, M., Galdi, A., Foltis, P., Pillai, S. (2010). Complementary effects of antioxidants and sunscreens in reducing UV-induced skin damage as demonstrated by skin biomarker expression. *J. Cosmet. Laser Ther.* [Epub. ahead of print].

Paltridge, G. W., Barton, I. J. (1978). Erythemal ultraviolet radiation distribution over Australia-the calculations, detailed results and input data Division of Atmospheric Physics Technical Paper 33 (Australia: Commonwealth Scientific and Industrial Research Organization).

Pandel, R., Poljšak, B., Godić, A., Dahmane, R. (2013). Skin photoaging and the role of antioxidants in its prevention. *ISRN Dermatology (Online)*, ISSN 2090-4606, 2013, vol. 2013, 1-11.

Pattison, D. I., Davies, M. J. (2006). Actions of ultraviolet light on cellular structures. *EXS.*, (96), 131-57.

Poljsak, B., Pesti, M., Jamnik, P., Raspor, P. (2011). Impact of environmental pollutants on oxidation-reduction processes in the cell environment. In: Dr. Jerome Nriagu (Ed.). *Encyclopedia of Environmental Health*. Elsevier.

Quan, T., He, T., Kang, S., Voorhees, J. J., Fisher, G. J. (2002). Ultraviolet irradiation alters transforming growth factor beta/smad pathway in human skin in vivo. *Journal of Investigative Dermatology,* 119(2), 499-506.

Rasche, C., Elsner, P. (2010). Skin aging: A brief summary of characteristic changes. In: *Textbook of aging skin*. Ferage, M. A., Miller, K. W., Maibach, H. I. (Eds.). Berlin: Springer-Verlag.

Rocquet, C., Bonté, F. (2002). Molecular aspects of skin ageing-recent data. Acta dermatologica Alpina, Pannonica ed. *Adriatica*, 11(3), 71-94.

Salinaro, A., Emeline, A. V., Zhao, J., Hidaka, H., Ryabchuk, V. K., Serpone, N. (1999). *Pure Appl. Chem.*, 71, 321.

Sander, C. S., Chang, H., Salzmann, S., Müller, C. S., Ekanayake-Mudiyanselage, S., Elsner, P., Thiele, J. J. (2002). Photoaging is associated with protein oxidation in human skin in vivo. *J. Invest. Dermatol.*, 118(4), 618-25.

Santos, J. M., Miranda, M. S., Esteves da Silva, J. C. G. (2012). *Water Research,* 46, 3167-76.

Serpone, N., Dondi, D., Albini, A. (2007). *Inorganica Chim. Acta*, 360, 794.

Scharffetter-Kochanek, K., Brenneisen, P., Wenk, J., Herrmann, G., Ma, W., Kuhr, L., Meewes, C., Wlaschek, M. (2000). Photoaging of the skin from phenotype to mechanisms. *Exp. Gerontol.*, 35(3), 307-16.

Scharffetter-Kochanek, K., Wlaschek, M., Brenneisen, P., Schauen, M., Blaudschun, R., Wenk, J. (1997). UV-induced reactive oxygen species in photocarcinogenesis and photoaging. *Biol. Chem.*, 378(11), 1247-57.

Schlumpf, M., Kypke, K., Wittassek, M., Angerer, J., Mascher, H., Mascher, D., Vökt, C., Birchler, M., Lichtensteiger, W. (2010). *Chemosphere*, 81 (10), 1171-83.

Schmid, D., Muggli, R., Zülli, F. (2002). Collagen glycation and skin aging. *Cosm. Toil.*, 118-24.

Seo, J. Y., Lee, S. H., Youn, C. S., Choi, H. R., Rhie, G., Cho, K., Kim, K. H., Park, C., Eun, H. C., Chung, J. H. (2001). Ultraviolet radiation increases tropoelastin mRNA expression in the epidermis of human skin in vivo. *J. Invest. Dermatol.*, 116, 915-9.

Serri, R., Iorizzo, M. (2008). Cosmeceuticals: focus on topical retinoids in photoaging. *Clin. Dermatol.*, 26(6), 633-5.

SORG (1990). *United Kingdom Stratospheric Ozone Review Group*. Third Report. London: HMSO.

Talwar, H. S., Griffiths, C. E., Fisher, G. J., Hamilton, T. A., Voorhees, J. J. (1995). Reduced type I and type III procollagens in photodamaged adult human skin. *J. Invest. Dermatol.*, 105(2), 285-90.

Trifunovic, A., Wredenberg, A., Falkenberg, M., et al. (2004). Premature ageing in mice expressing defective mitochondrial DNa polymerase. *Nature*, 429, 417-23.

Tsuji, N., Moriwaki, S., Suzuki, S., Takema, Y., Imokawa, G. (2001). The role of elastases secreted by fibroblasts in wrinkle formation: implication through selective inhibition of elastase activity. *Photochem. Photobiol.*, 74, 283-90.

Tsukahara, K., Takema, Y., Moriwaki, S., Tsuji, N., Suzuki, Y., Fujimura, T., Imokawa, G. (2001). Selective inhibition of skin fibroblast elastase elicits a concentration-dependent prevention of ultraviolet B-induced wrinkle formation. *J. Invest. Dermatol.*, 117, 671-7.

Tyrrell, R. M. (1991). UVA (320–380 nm) radiation as an oxidative stress. *Oxidative Stress: Oxidants and Antioxidants*. In: Sies, H., ed. New York: Academic Press, 57-78.

Varani, J., Warner, R. L., Gharaee-Kermani, M., Phan, S. H., Kang, S., Chung, J. H., Wang, Z. Q., Datta, S. C., Fisher, G. J., Voorhees, J. J. (2000). Vitamin A antagonizes decreased cell growth and elevated collagen-degrading matrix metalloproteinases and stimulates collagen accumulation in naturally aged human skin. *J. Invest. Dermatol.*, 114(3), 480-6.

Vidal, L., Chisvert, A., Canals, A., Salvador, A. (2010). *Talanta* 81(1-2), 549-55.

Watson, R., Griffiths, E. M., Craven, N. M., Shuttleworth, A., Kielty, C. M. (1999). Fibrillin-rich microfibrils are reduced in photoaged skin. Distribution at the dermal-epidermal junction, *J. Invest. Dermatol.*, 112, 782-7.

Yaar, M., Gilchrest, B. A. (2003). Aging of skin. In: Freedberg, I. M., Eiser, A. Z., Wolff, K., et al. (Eds.). *Fitzpatrick's dermatology in general medicine*, vol. 2. New York: McGraw-Hill.

Yaar, M., Eller, M. S., Gilchrest, B. A. (2002). Fifty years of skin aging. *J. Invest. Symp. Proc.*, 7, 51-58.

Yano, K., Ouira, H., Detmar, M. (2002). Targeted overexpression of the angiogenesis inhibitor thrombospondin-1 in the epidermis of transgenic mice prevents UV B-induced angiogenesis and cutaneous photodamage. *J. Invest. Dermatol.*, 118, 800-5.

Yarosh, D. B. (2010). DNA damage and repair in skin aging. In: *Textbook of aging skin*. Farage, M. A., Miller, K. W., Maibach, H. I. (Eds.). Berlin: Springer-Verlag.

Yeates, D. B., Mauderly, J. L. (2001). *Environ. Health Perspect.* 109(4), 479-81.

Zhuang, R., Žabar, R., Grbović, G., Dolenc, D., Yao, J., Tišler, T., Trebše, P. (2013). *Acta Chimica Slovenica*, 60(4), 826-32.

Zucchi, S., Blüthgen, N., Ieronimo, A., Fent, K. (2011). *Toxicology and Applied Pharmacology*, 250(2), 137-46.

In: Sun Exposure
Editor: Collin P. Raines
ISBN: 978-1-63482-087-5

Chapter II

Photoprotection Practices

Jacqueline Selph, MD, Ritva Vyas, MBChB and Meg Gerstenblith, MD
University Hospitals/Case Medical Center, Cleveland, OH, US

Abstract

Sun protection practices play a prominent role in dermatology in terms of photoaging, photocarcinogensis, and photodermatoses. In this chapter, we outline the various methods of sun protection. Naturally occurring protective agents, such as geographic and environmental variations create a substantial difference in exposure to ultraviolet (UV) radiation. Physical protective barriers including glass, clothing, and sunglasses can provide UV protection, but there is great variability in their defensive properties. Sunscreens play a major role in sun protection practices and vary substantially in chemical composition and effectiveness regarding protection against UVA versus UVB radiation. We will discuss inorganic products versus organic products in sunscreens as well as their associated advantages and disadvantages. Finally, systemic agents, such as plant extracts, carotenoids, polyphenols, and other antioxidants may prove to be promising adjuncts for sun protection in the future.

Introduction

The electromagnetic spectrum emitted from the sun includes x-rays, ultraviolet radiation (UVR), visible light, infrared, and microwaves/ radiowaves. UVR, which includes ultraviolet A (UVA 320-400nm), ultraviolet B (UVB 290nm-320nm), and ultraviolet C (UVC 200-290nm) has a variety of negative effects on the skin. In addition to being a major environmental risk factor for melanoma and non-melanoma skin cancers, it also contributes to skin aging, photosensitivity in certain dermatologic conditions, and immunosuppression. UVB is the major type of UVR known to cause direct photochemical damage through the formation of pyrimidine dimers and subsequent gene mutations. UVA can be further subdivided into UVA2 (320-340 nm) and UVA1 (340-400nm). UVA has indirect effects on photocarcinogenesis via the formation of reactive oxygen species (ROS).

Furthermore, evidence suggests that UVA may have a more damaging role compared to UVB in long-term sun damage. In this chapter, we will outline the various forms of photoprotection including environmental protections, intrinsic properties of the skin, and physical, topical and systemic agents.

Environmental Protections

Atmosphere and Geography

The ozone (0_3) layer, present in the stratosphere at 10km to 50km above the surface of the earth, absorbs high quantities of short UVB and UVC radiation but very little UVA. However, the ozone layer is not uniform in thickness, with the concentration increasing toward polar regions. Chlorofluorocarbons used as aerosol propellants have led to ozone depletion, especially towards the South Pole. Sunburns and photosensitivity disorders have been shown to increase after acute, sudden episodes of UVB radiation second to ozone depletion (Norval et al., 2011). Ozone depleting substances have been phased out, and therefore a slow recovery of ozone is anticipated over the coming decades (Bais et al., 2014).

Scattering by nitrogen and oxygen molecules also helps to deflect UVC resulting in near complete attenuation of the UVC reaching the earth's surface. The atmosphere allows about 10% of UVB light and almost all UVA and visible light to reach the earth's surface; therefore, on a summer day >95% of UVR reaching the earth's surface is UVA, while only 4% is UVB. UVR exposure is greatest at the equator and high altitudes. For every degree increase in latitude away from the equator, there is a 3% decrease in the transmission of UVB, and for every 300 meter increase in elevation, there is an approximately 4% increase in the intensity of UVR, the majority of which is UVB (Rigel, Rigel, and Rigel, 1999).

Time of Day and Season

At the solar zenith, the path of UVR through the ozone is the shortest; thus, substantially less UVR is absorbed. UVA light penetrates without absorption through the atmosphere; therefore, its level is constant through daylight hours. However, UVB light varies with atmospheric absorption; it is strongest from 10 AM to 2 PM, when the path of transmission is shortest. An observational study in Denmark showed that 50% of the total daily UVR dose reaches the earth's surface between 12 PM and 3 PM (Thieden, Philipsen, Heydenreich, and Wulf, 2004). Likewise, UVR is strongest in the summer because of the elliptical orbit of the sun (Diffey, 2002).

Clouds, Particulates, and Reflections

Clouds, fog, and haze are estimated to reduce ultraviolet levels between 10%-90%. The US national weather service calculates that overcast skies allow only 31% of UV transmission, broken clouds 73%, and scattered clouds 89%; however, very heavy cloud cover (i.e. storm clouds), can virtually eliminate ultraviolet exposure (US Environmental Protection Agency, 2012). While clouds reduce the UVR intensity, the infrared (heat) intensity is reduced to a much greater extent. Without the warning sensation of warmth from infrared light, the risk of overexposure to UVR greatly increases on cloudy days.

Pollutants such as soot, nitrogen dioxide, and sulfur dioxide act similarly to clouds, reducing UVR through scattering. Shorter wavelengths are scattered to a greater extent than longer wavelengths. Therefore, in large urban areas, UV irradiance is reduced compared to more underdeveloped areas (Mckenzie et al., 2008).

The reflection of UVR off various surfaces also greatly contributes to overall UVR exposure. Snow, sand, and metal reflect up to 90% of UVR; almost doubling the UVR exposure of skiers and beachgoers. However, most surfaces reflect less than 10% of ultraviolet light.

While little reflection occurs on still water, UVR can penetrate to a depth of one meter, exposing swimmers to substantial radiation Seawater, due to its motion and relatively high particulate count, can reflect up to 15% of UVR (Kromann, Wulf, Eriksen, and Brodthagen, 1986; Lautenschlager, Wulf, and Pittelkow, 2007).

Shade

Although data have shown "shade seekers" to be relatively protected from the harms of UVR, there is still substantial exposure to UVA light in the shade (Turnbull and Parisi, 2003). Approximately 50% of all exposure to UVA light occurs in the shade. While the sun protection factor (SPF) of single trees can be as low as 4, dense foliage can increase this beyond SPF 50, reducing UVR by as much as 95% (Moise and Aynsley, 1999).

An average beach umbrella (not made specifically for UV protection) offers almost no UVR protection. Newer sun protection umbrellas can provide increased SPF from overhead; however, with typical beach use, much UVR is reflected from sand and water (Thieden et al., 2004).

Solar Ultraviolet Index

The global solar ultraviolet index (UVI), developed by the World Health Organization (WHO) in consortium with various international agencies, takes into account the above mentioned environmental elements and provides a numerical guide for the level of solar UVR at any given location and date as well as how much protection is recommended.

The UVI ranges from 1 to 11+ with levels 1-2 requiring no additional protection, levels 3-6 requiring protection, and levels 8 and above requiring extra protection (World Health Organization, 2014).

Intrinsic Photoprotective Properties of the Skin

The main human protective barrier to UVR is the skin. However, while the skin serves as a protective barrier for the internal organs, it absorbs most UVR, sometimes leading to its own damage. UVB light is primarily absorbed in the epidermis, but UVA light can penetrate deep into the dermis. As early as half an hour following UVR exposure, apoptotic keratinocytes, known as sunburn cells, can be observed histologically. This apoptotic process is a protective mechanism to rid the body of cells with the potential for malignancy second to DNA damage (Baron and Suggs, 2014).

Chromophores are defined as molecules that absorb light-energy and in turn give off a color of their own. In the body, the DNA bases purine and pyrimidine act as chromophores and absorb most of the effects of UVB, which can lead to the development of cyclobutane and pyrimidine dimers. Such damage, if missed by repair mechanisms, can lead to mutations or cytotoxicity.

The aromatic amino acids, tryptophan and tyrosine, have significant absorption in the UVB range, but little UVA absorption. Proteins can photobind to DNA, and there is some evidence that photoactivation of a protein may be an important step in transcription factor up regulation (Young, 1997).

Melanins are another important chromophore in the skin. Melanin accumulates within keratinocytes and melanocytes in the perinuclear area and functions as a "cap" to shield DNA from UVR, absorbing 50%-75% of UV rays; additionally, melanin acts as a free radical scavenger, antioxidant, and superoxide dismutase that reduces ROS (Brenner and Hearing, 2008).

Eumelanins are brown to black nitrogenous pigments formed from the oxidative polymerization of 5,6 dihydroxyindoles (DHIs), whereas pheomelanins are alkali-soluble yellow to reddish-brown pigments formed from the oxidative polymerization of cysteinyldopas. Pheomelanins, seen in red haired/fair skinned individuals, are formed from loss of function polymorphisms in the melanocortin 1 receptor gene (*MC1R*), of which greater than 100 have been identified (Dessinioti, Antoniou, Katsambas, and Stratigos, 2011; Gerstenblith, Goldstein, Fargnoli, Peris, and Landi, 2007). Pheomelanin has weak UV shielding ability compared to eumelanin and also perpetuates damaging ROS.

There is evidence that loss of function polymorphisms in *MC1R* are associated with an increased risk of melanoma, indicating a superior protective role of eumelanin over pheomelanin (Pasquali et al., 2015; Rees, 2000).

Recently, data in mice have suggested that even in the absence of UVR, pheomelanin contributes to melanoma carcinogenesis through reactive oxidative damage (Mitra et al., 2012). More darkly pigmented skin, containing mostly eumelanin, is less susceptible to the damaging effects of UVR than lightly pigmented skin, which in large part explains the increase in melanoma, squamous cell carcinoma, and basal cell carcinoma observed in fair skinned individuals.

Other photoprotective agents in the skin include heme and porphyrin. Oxyhemoglobin and reduced hemoglobin can absorb bands in the UVA/UVB range, as well as blue, green, and yellow visible light. Porphyrins absorb light generally between 400nm-410 nm, and generate ROS upon exposure. Unfortunately, these molecules accumulate to high levels in cutaneous porphyrias causing photosensitivity (Young, 1997).

Physical Photoprotective Agents

Glass

Although eighty percent of the average day of Americans is spent indoors, contemporary design increasingly incorporates many large window areas. It has been estimated that individuals who work indoors receive on average 8.5 standard erythema doses (SEDs) per day of UVR in spring months and at least 2 per day in winter months (Parisi et al., 2000). Additionally, a large number of Americans commute, with the average American spending between 80-90 minutes per day in automobiles. In studies of UVR exposure in cars, an increased prevalence of facial photodamage, actinic keratosis, melanoma, and non-melanoma skin cancers were seen on the driver's exposed side (Butlers ST, 2010; Singer, Hamilton, Voorhees, and Griffiths, 1994).

The majority of commercially used glass is soda lime glass, made up of a mixture of silica, salt cake, limestone, dolomite, feld-spar, soda ash, and typically recycled broken glass. Through melting and slow cooling, the glass develops a random, disorganized non-crystalline structure, which on its own provides little UVR protection (Tuchinda, Srivannaboon, and Lim, 2006).

The main types of glass are clear glass, tinted (heat-absorbing) glass, reflective glass, low-emissivity glass, laminated glass, UVR-blocking coated glass, and insulating glass. Tinted glass may absorb 40%-50% of incoming solar energy and has less UVR and visible light transmission compared to clear glass.

Reflective glass uses a metal oxide coating to give glass a mirror-like appearance, which helps to minimize unwanted solar heat gain and reduces UVR and visible light transmission (Almutawa, Vandal, Wang, and Lim, 2013). Low Emissivity (Low-E) glass has a surface coating of microscopically thin transparent layers of silver sandwiched between antireflective metal oxide coatings. It significantly reduces the loss of generated heat, and may decrease UVR transmission from 60% down to 20%; however, UVA is largely unimpeded by this coating (National Glass Association, 2014). Laminated glass is a combination of two pieces of glass bonded with a tough plastic interlayer. The main benefit is to prevent injury because if broken, large fragments do not fall free; however, it also filters more than 99% of UVR without sacrificing visible light transmission. UVR blocking coated glass blocks more than 98% of UVR. Finally, insulated glass combines Low-E glass with UVR-blocking coated glass to block more than 99% of UVR transmission while eliminating up to 70% of unwanted solar heat gain (Almutawa et al., 2013; Tuchinda et al., 2006).

Automobile and Airplane Glass

Because of the hazards of broken glass, all automobile windshields are made of laminated glass, which filters UVB and most UVA radiation below 380 nm. However, side, rear, and overhead windows are usually made from tempered glass, which block all UVB radiation but only 21% of UVA radiation. Window tinting can help to further decrease this exposure; however, it is generally only permitted on rear windows. A subject sitting near the driver's side window of non-laminated glass could be exposed to a 5-J/cm2 dose of UVA in 30 minutes; enough to produce an eruption in patients with polymorphous light eruption (Hampton, Farr, Diffey, and Lloyd, 2004). In 2011, the National Highway Traffic Safety Administration mandated stricter requirements to mitigate side and rear window passenger ejections. As a result, more side and rear windows are being made with laminated glass.

Window films can be applied to side and rear windows, reducing transmission of visible light and infrared radiation while decreasing interior heat gain and minimize fading. Most films are made of multiple layers of polyethylene terepththalate (PET), a polyester resin, which gives only a small reduction in visible light (Almutawa et al., 2013).

However, infrared radiation exposure can contribute to skin aging by decreasing pro-collagen and increasing cytokines. The primary objective of UVR blocking is accomplished by adding UVR absorbers or adhesives to the films.

These films must comply with federal and state standards, which mandate no higher than a 30% blockade of visible light. However, as more automobiles manufacturers use laminated glass for side and rear windows, window films may become unnecessary for UVR protection purposes.

Airplane windshields are commonly made of polycarbonate plastic or laminated glass, which block >99% of UVB. Plastic offers better UVA protection compared to glass; however UVA transmission as high as 53% has been reported (Nakagawara, Montgomery, and Marshall, 2007). A recent study demonstrated that pilots flying at 30,000 feet for 56 minutes were exposed to the same UVA radiation dose as a 20 minute tanning bed session, which may contribute to carcinogenesis and the increase in melanoma observed in pilots and cabin crew (Sanlorenzo et al., 2014). Because of the increase in UVR at high altitudes, future recommendations may mandate UVR absorbing films on airplane windshields.

Architectural Glass for Buildings

All types of glass, as previously mentioned, can be used as architectural glass. The highest transmission of UVA is through smooth annealed glass (74.3% of UVA) followed by tempered glass (71.6%), textured annealed glass (44.6%), and finally laminated glass, which allows <1% UVR through (Tuchinda et al., 2006). These values can vary based on the thickness and color of the glass. Just as with the tinting of automobile glass, many architectural buildings apply energy efficient glaze, which reduces heat gain and loss as well as some UVR protection (Almutawa et al., 2013).

Sunglasses

Several eye disorders are related to sun exposure, such as cataracts, macular degeneration, pterygium, and keratitis. UVR protection through sunglasses may help to reduce the incidence of these conditions. Australia leads the way in sunglass guidelines, establishing the first national standard in 1971. Currently, there are three available national standards for sunglasses, the

American Standard (ANSI Z80.3), the European standard (EN 1836:2005), and the Australian/New Zealand standard (AS/NZS 1067:2003).

The European and Australia/New Zealand standard are very similar, utilizing five lens categories (from very light tint to very dark tint), with the main differences being the definition of UVA (315-380 nm per EN and 315-400 nm per AS) and that European standards allow up to 10% of UVB (defined as 280-315 nm), while Australian standards allow a maximum of 5% UVB (Dain et al., 2010). The American standards have only four categories of lens tint but are further differentiated based on use for cosmetic or special purposes.

Furthermore, the American standards differ depending on whether the lens is to be used for normal (i.e. commuting from home to work) or prolonged use. A darkly colored special purpose lens should transmit 1% or less of UVB and 50% or less of UVA; however, a commonly sold cosmetic lens for normal use can allow up to 12.5% UVB and all UVA light through (Almutawa and Buabbas, 2014).

Table 1. Common types of glass used in residential and commercial buildings

Types of Glass	Comment
Clear glass	Protection from outside elements while allowing visible light transmission
Tinted (heat-absorbing) glass	Absorbs solar energy to ↓ heat gain Some ↓ in visible light and UVR
Reflective glass	Reflects heat and visible light Moderate ↓ in visible light and UVR
Low-emissivity glass	Significant ↓in heat loss and ↓ heat gain Allows full visible light, ↓ UVB
Laminated glass	>99% UVB blocked no ↓ in visible light Does not fragment when broken
UV blocking coated glass	Allows all visible light blocks >98% UVR
UV blocking coated insulated glass	Combines Low-E glass with UV coated glass to ↓both UVR and solar heat gain

Compliance with these standards in the United States is voluntary, whereas compliance with the European and Australian/New Zealand standards is mandatory. In Europe, quality assurance to standards can be done by the manufacturer themselves, but AS/NZ mandate inspection by an independent party (Almutawa et al., 2013). In a recent study testing European sunglasses that comply with standards, 17% failed to meet these standards when assessed by an independent party (Dain et al., 2010).

In addition to lens standards, the size and geometry of sunglasses make a significant difference in UVR exposure.

Table 2. Summary of Australian, European and United States Sunglass Standards

		0 (very light tint)	1 (light tint)	2 (medium tint)	3 (dark tint)	4 (very dark tint)
UVB	EN (280-315 nm)	10% LT	10% LT	10% LT	10% LT	10% LT
	AS (280-315 nm)	5% LT	5% LT	5% LT	5% LT	5% LT
	US (280-315 nm) Normal Use	n/a	12.5% LT	12.5% LT	1% LT	1% LT
	US (280-315 nm) Prolonged Use	n/a	1% LT	1% LT	1% LT	1% LT
UVA	EN (315-380 nm)	LT	LT	LT	50% LT	50% LT
	AS (315-400 nm)	LT	LT	LT	50% LT	50% LT
	US (215-380 nm) Normal Use	n/a	LT	LT	50% LT	50% LT
	US (215-380 nm) Prolonged Use	n/a	50% LT	50% LT	50% LT	50% LT

*LT = Luminous Transmittance (the fraction of incident light that passes through the sample).

Maximum UVR to the eye occurs when solar radiation is parallel to the eye; however, radiation from above or below the eye also makes a sizeable contribution. The best protection is achieved with wraparound sunglasses or side shields (Almutawa and Buabbas, 2014). Unfortunately, UVR from other angles can reflect off the inner surface of the lens increasing exposure to the eye. Australia is the only country with a standard for lens size, which mandates 28 mm for adults and 24 mm for children (Almutawa et al., 2013).

UVR eye protection is extremely important in young children because their ocular lenses do not filter UVR as well as developed adult lenses. Visible light and UVR are able to reach the retina in young children because of their large pupillary size, increasing the risk for macular degeneration (Rosenthal, Bakalian, Lou, and Taylor, 1988).

While dark lenses often provide more UVR protection, they also result in pupil dilation, which could increase retinal UVR exposure.

Other Eye Protection

While sunglasses can be an effective method of UVR eye protection, they also reduce the squint mechanism, which naturally allows less radiation to reach the eye. Newer contact lenses incorporate some UVR-blocking properties, which helps to mitigate this effect. The Federal Drug Administration (FDA) mandates that Class I contact lenses must block more than 90% of UVA (316-380 nm wavelength) and 99% of UVB (280-315 nm), whereas Class II contact lenses, which are intended for general purposes, must block more than 70% of UVA and 95% of UVB (Walsh and Bergmanson, 2011).

Polycarbonate is a new thermoplastic material first used in the aerospace industry that is now being transitioned to use in glasses. Studies of airplane windshields found that polycarbonate transmitted almost no UVR below 380 nm, therefore blocking all UVB and almost all UVA. This material is lightweight, strong, and resistant to breakage. Newer variations of traditional polycarbonate claim to block almost 100% of UVR, and offer an impact resistant, UVR protective alternative to traditional lenses (Dain, 2012).

Photoprotective Clothing

Over the past twenty years, photoprotective fabric has been developed as a reliable means of decreasing exposure to UVR. First coined in Australia in 1996, the term "UV protection factor" (UPF) is used to describe the protection afforded by textiles, which utilizes spectrophotometric methods to measure transmission of both UVA and UVB in vitro. The UPF is calculated by combining these transmission data with erythema effectiveness. Measuring protection of fabrics in vivo is done through exposure to incremental UVB doses on protected and unprotected skin. The minimal erythema dose (MED) for protected *vs.* unprotected skin is measured, and a protection factor is calculated. There are some reports of good correlation between in vitro and in vivo measurements; however, in vitro measurements are currently the standard. Unfortunately, not all clothing is equal in photoprotection. One-third of summer clothing has a UPF of less than fifteen (Gambichler, Altmeyer, and Hoffmann, 2002) and the average protection afforded by a light-colored cotton shirt was only UPF 10 (Wright, Hart, and Peirce, 1998). Tightly woven fabrics, dark colors, wool, and polyester all provide increased UVR protection; however, they are commonly worn in winter months when solar radiation is diminished (Gambichler et al., 2002). Cotton, linen, acetate, and rayon generally have a UPF less than 15, whereas thick denim provides a UPF of 1700. Additives such as Tinosorb FD (BASF, Basel, Switzerland), contained in the product "Sun Guard®" by Rit®, absorb UVR significantly decreasing transmission and offer a UPF up to 30. Products like this claim to last up to 20 washes, but the type of detergent used, the weave of the fabric, and bleaching can significantly decrease longevity.

Hydration can also have a significant effect on the UPF of clothing. When saturated, linen, viscose, and polyester significantly increase their UPF, making them ideal for beach or swim wear. However, cotton and polyester fabrics show a significant decrease in UPF when saturated. It is estimated that a wet cotton shirt provides a UPF of only 3-4 (Gambichler et al., 2002; Gambichler, Hatch, Avermaete, Altmeyer, and Hoffmann, 2002).

Hats

Hats can be problematic for sun protection because they shade the face, decreasing the infrared exposure and the feeling of heat, while still allowing significant UVR exposure. A wide-brimmed hat (>7.5 cm) may only provide an SPF of 7 for the nose, 3 for cheeks, 5 for the neck, and 2 for the chin; this is due to reflection of UVR off surfaces below the brim of the hat. A narrow brim hat provides no more than SPF 1.5 for the nose and minimal protection for other areas (Diffey and Cheeseman, 1992). Additionally, the weave of the hat can make a significant difference, with a loosely woven straw hat offering little UVR protection but still significant shade to the face (Jansen, Wang, Burnett, Osterwalder, and Lim, 2013).

Topical Photoprotective Agents

Makeup

Most commercial brands of foundation now contain sunscreen, which offers significant photoprotection and will be discussed subsequently.

Foundation makeup without sunscreen still provides an SPF of 3 to 4 as a result of pigment in the foundation. However, makeup can give a false sense of protection, as it is only applied once per day and loses its protective properties within a few hours. Makeup also can migrate into dermatoglyphs and accumulate in follicles, decreasing its photoprotective properties. This process happens even more rapidly in situations with increased perspiration, sebum production, or tearing. Dermatologists recommend using longwearing sunscreen underneath foundation, even if makeup contains no SPF to increase overall photoprotection.

Sunscreens

The first commercial sunscreen was introduced in 1928; since that time, sunscreen has become an essential part of sun protection practices. The first commercial sunscreen, an oil preparation containing benzyl salicylate, was marketed in 1936 by the future founder of L'Oreal, although it was in the 1970s that the concept of sun protection factor was introduced broadly, creating a comparable market for sunscreens (Jansen, Osterwalder, Wang, Burnett, and Lim, 2013).

Currently, there are only 18 approved agents in the FDA monograph, in contrast to 28 different sunscreens in the Europe Union and at least 34 approved in Australia. Since 1978, the FDA has only approved the addition of three compounds – avobenzone, zinc oxide, and the more recently approved ecamsule (Food and Drug Administration, 1999). The lag between the United States and other countries lies in the fact that sunscreens in the US are treated as over the counter medications, which necessitate a more rigorous investigation than handling it as a cosmetic product as it is treated in Europe and Australia.

The Sun Protection Factor (SPF) measure is used to evaluate the efficacy of topical photoprotectants. Unlike the measurement done for clothing (UPF), SPF is measured in vivo by testing the MED of volunteers with Fitzpatrick type I, II, or III skin. MED, also known as the sunburn threshold, is the minimal UVR dose, specifically UVB, since it is the primary erythema inducer, required to produce a faint pink response in the skin (Baron and Suggs, 2014). The UVR dose is calculated:

$$Dose\ (mJ/cm^2) = Irradiance\ (mJ/s \bullet cm^2)\ x\ exposure\ time\ (s)$$

To measure the MED, adjacent areas of skin are exposed to increasing amounts of UVR and then 24 hours later, the MED is defined as the amount of UVR that produced visually apparent erythema. The testing of all sunscreens uses the same density of 2 mg/cm and measures the MED on treated versus untreated skin (Schalka and Silva dos Reis, 2011). SPF is then calculated:

$$SPF = MED\ (protected\ skin)\ /\ MED\ (unprotected\ skin)$$

It is important to note that this testing takes place under high intensity solar simulators, which can cause an erythema dose in as little as two minutes, compared to much less intense natural sunlight. Therefore, the SPF is not a measure of duration of UV exposure (Baron and Suggs, 2014; Schalka and Silva dos Reis, 2011). Additionally, the erythema dose is a combination of UVB and UVA2, but because UVA1 does not cause erythema, it is not regularly represented in an SPF value (Jansen et al., 2013).

Different countries have imposed differing regulations for UVA testing. Japan currently uses persistent pigment darkening (PPD) as a clinical endpoint of UVA protection, and rates sunscreens as PA+ to PA++++ (Japan Cosmetic Industry Association, 1995). The EU also uses PPD as testing, but simply mandates that all marketed sunscreens contain UVA protection at least one-third of the labeled SPF (European Cosmetic Toiletry and Perfumery Association, 2011). In 2011, the US FDA mandated the use of testing for UVA protection through an in vitro critical wavelength (CW) testing. This test uses a solar simulator to deliver four-times the MED dose (on a Fitzpatrick type II skin) to the test product, and then measures transmittances from 290-400 nm (the UVA range). CW is defined as the wavelength at which 90% of the area under the absorbance curve occurs. Sunscreens are then simply classified as broad spectrum if their critical wavelength is ≥ 370 nm (Jansen et al., 2013).

Organic Sunscreen Agents

Organic sunscreen agents primarily absorb UVR through chromophores with a conjugated π-electron system. Generally, the larger the molecule, the more conjugated double bonds are present, which shifts the absorption toward longer wavelengths. Thus, smaller molecules are more suitable as UVB filters and larger molecules as UVA filters (Sambandan and Ratner, 2011). Currently, all organic UVR absorbers are aromatic compounds with multiple conjugated π-electron systems. Photostability is also an important property of organic agents. If the energy absorbed is not dissipated quickly into heat, it can lead to degradation of the UVR absorbers.

The formation of a reversible isomer (tautomerization) is used in the menthyl anthranilate molecule through use of the orthoamino group and in bemotrizinol and bisctrizole through an orthohydroxy group (Gaspar and Maia Campos, 2006).

Organic agents can be further subdivided into UVA and UVB absorbers. One of the first widely available and most potent UVB protectors is para-aminobenzoic acid (PABA); however, it has multiple disadvantages. In addition to staining clothing, it is a common contact and photoallergen and was found to be a potent carcinogen in vitro, although the in vivo significance is unknown. It has been largely replaced by the less potent but more tolerable Padimate O (Jansen et al., 2013). Cinnamates, such as octinoxate and the less commonly used cinoxate, are less potent UVB absorbers than PABA derivatives but are well-tolerated and rarely cause irritation. While they do not stain, they also have reduced water resistance, making frequent reapplication necessary. Salicyclates, such as octisalate, homosalate, and trolamine salicylate, are the weakest UVB agents but are often used in combination with each other or other UVR filters to augment UVB protection. In particular, octisalate and homosalate are highly photostable and when used in combination with other UVR filters, help reduce photodegradation (Sambandan and Ratner, 2011). The newest UVB absorber, ensulizole, provides a lighter and less oily consistency than other organic agents, and therefore is becoming popular for facial wear (Sambandan and Ratner, 2011).

The first FDA-approved organic UVA1 filter was avobenzone (butyl methoxydibenzoylmethane), which filters most UVA radiation. Its effectiveness is limited because it is extremely photolabile, and its protective properties decrease >50% within an hour of use. Unfortunately, it can also affect the stability of other sunscreen agents; therefore, much effort has surrounded stabilizing formulas (Palm and O'Donoghue, 2007). One of the most prominent is the Helioplex® stabilizing technology from Neutrogena®, which combines avobenzone, oxybenzone, and diethylhexyl 2,6 napthalate (Jansen et al., 2013).

Ecamsule is the most recently FDA approved agent, which broadly filters UVA but not UVB. It is therefore commonly combined with avobenzone and octocrylene. Unlike avobenzone, it is very stable and not subject to photodegradation. Initially, it was only available as Anthelios® in La Roche-Posay® products, but it has now been incorporated into other L'Oreal® products as well (Fourtanier, Moyal, and Seite, 2008). Lastly, meradimate, a weaker UVA2 filter, is in combination with other agents to provide increased UVA protection (Sambandan and Ratner, 2011).

The most common broad-spectrum UVA and UVB organic agents are in the class of benzophenones. Currently, oxybenzone, sulisobenzone, and dioxybenzone are approved in the US, with oxybenzone being the most widely used. Recently, this class has received attention because of an increase in contact allergies. In addition to being used as sunscreens, they are also added to many colored personal use products to prevent color degradation; therefore, it is estimated that 96% of the US population has been exposed to a benzophenone (Heurung, Raju, and Warshaw, 2014).

Additionally, oxybenzone has been shown in vitro to produce estrogenic and antiandrogenic effects. In an in vivo study of female rats exposed to oxybenzone, the uterus size was 23% greater than controls; however, the dosage used far exceeded the normal level of human exposure (Schlumpf et al., 2001). While overall the evidence is inconclusive, there is still concern regarding benzophenone's use.

Inorganic Agents

In contrast to organic agents, inorganic agents (previously known as physical filters) utilize a film of inert metal particles to form an opaque barrier, which reflects and scatters UV light. The two widely used inorganic agents are zinc oxide, which offers predominantly UVA protection, and titanium dioxide, which provides predominantly UVB protection (Sambandan and Ratner, 2011). Inorganic agents were initially unpopular because they required a thick, opaque application, which was aesthetically displeasing.

Newer formulations have been able to microsize both titanium dioxide and zinc oxide, reducing the particle size from 200nm-500 nm to 10nm-50 nm, improving aesthetics (Pinnell, Fairhurst, Gillies, Mitchnick, and Kollias, 2000). Inorganic agents are not susceptible to photodegradation, so they provide less variability in photoprotection.

Inorganic sunscreens are recommended for children and sensitive individuals because they have no known skin irritating or sensitizing potential. Additionally, children, especially infants, have a higher surface area to volume ratio compared to adults, which causes concern for absorption of topically applied medications. While the absorption potential of organic agents such as benzophenone is controversial, inorganic agents have been shown repeatedly to have no percutaneous penetration (Jansen et al., 2013).

Therefore, in children under two years of age, only inorganic sunscreens should be applied, and in children less than six months, other photoprotective methods are recommended (Council on Environmental Health, Section on, and Balk, 2011).

Table 3. FDA approved sunscreen agents

Class	FDA approved ingredient	Peak Absorption Wavelength (nm)	Range of Protection (nm)	Note
Inorganic Agents				
	Titanium Dioxide	varies	290 – 350	Better UVB protection
	Zinc Oxide	varies	290 – 400	Better UVA protection
Organic UVB				
PABA derivatives	PABA	283	260 – 313	Stains clothing, contact allergen
	Padimate O	311	290 – 315	Largely replaced PABA
Cinnamates	Oxtinoxate	311	280 – 310	Decreased water resistance
	Cinoxate	311	270 – 328	
Salicylates	Octisalate	307	260 – 310	Photostable, reduce photodegredation of other agents
	Homosalte	306	270 – 328	
	Trolamine Salicylate	260-355	260 – 355	Water soluble, found in hair products
	Octycrylene	303	287 – 323	Costly, difficult to incorporate
	Ensulizole	310	290 – 340	Lightweight, less oily
Organic UVA				
	Avobenzone	360	310 – 400	Highly photolabile, in Helioplex®
	Meradimate	336	200 – 380	Weaker UVA2 filter
	Ecamsule	345	295 – 390	Broad spectrum, Anthelios®
Benzophenones	Oxybenzone	290, 325	270 – 350	Broad Spectrum, UVA and UVB protection
	Sulisobenzone	366	250 – 380	
	Dioxybenzone	352	206 – 380	

Systemic Photoprotective Agents

While currently the most popular form of photoprotection, topical agents have many limitations. The efficacy of topical agents is drastically reduced with improper application and reapplication, making a systemic alternative a logical route to pursue. Unfortunately, no current systemic therapies provide enough protection to be used as a sole agent; however, further research in this area is being pursued.

Polypodium Leucotomos

Polypodium leucotomos is perhaps the most well characterized photoprotective agent. As an extract from the fern leaf, it was used for centuries by Native Americans for its anti-inflammatory properties. Research on the compound has shown that it acts a scavenger to absorb free radicals and ROS, protects DNA by inhibiting the formation of pyrimidine dimers, increases the number of Langerhans cells in the skin, helps to inhibit mast cells from infiltrating into the skin, and overall reduces the effects of UVR (measured by an increase in the MED). In a study of 35 patients with long-standing polymorphous light eruption, daily administration of 480 mg to 1200 mg *P leucotomos* resulted in a significant proportion of patients becoming unresponsive to repeated UVA and UVB exposure (Tanew, Radakovic, Gonzalez, Venturini, and Calzavara-Pinton, 2012). In a larger study of high risk malignant melanoma patients, 1080 mg of *P leucotomos* increased the MED in all patients (decreasing UVR sensitivity) (Aguilera et al., 2013). The SPF of orally administered *P leucotomos* is estimated to be between 3-7, making it unacceptable as a sole photoprotective strategy, but an effective adjuvant therapy.

Carotenoids

Carotenoids are a class of micronutrients that act as antioxidants to provide skin protection by decreasing the free-radical induced damage to DNA. A class of vitamin A derivatives, they include lycopene, lutein, zeaxanthin, and betacarotene. Both betacarotene and lycopene have been show to decrease in skin concentration following UVR. In a large longitudinal study for 4.5 years, betacarotene supplementation was not shown to decrease non-melanoma skin cancer development (Green et al., 1999). However, other studies demonstrated that long term supplementation with betacarotene does provide protection from UVR-induced erythema (Heinrich et al., 2003).

Overall, the oral administration of carotenoids can provide some photoprotection, especially in comparison to topical application.

Afamelanotide

Afamelanotide, an α-melanocyte stimulating hormone analogue, induces epidermal melanin formation by binding to receptors on melanocytes, leading to increased melanocyte proliferation. A relatively new substance, it was granted investigational new drug status by the FDA in 2009.

Since that time, subcutaneous administration of the 13-amino acid analogue has been shown to increase tolerance to UVR exposure in patients with erythropoietic protoporphyria and solar urticaria (Harms, Lautenschlager, Minder, and Minder, 2009). Other α-melanocyte stimulating hormone analogues have been used outside of clinical or research settings by individuals seeking tan skin and weight loss.

These other analogues are less specific and interact with a wide range of receptor. There have been several case reports of α-melanocyte stimulating hormone analogues causing eruptive nevi or rapid changes to current nevi, therefore the FDA has issued warnings against the cyclic peptide (J. Harms, Lautenschlager, Minder, and Minder, 2009; Reid, Fitzgerald, Fabre, and Kirby, 2013).

Polyphenols

Polyphenols, most commonly phenolic acid, flavonoids, catechins, stilbenes, and proanthorcyanidins, have anti-oxidant, anti-inflammatory, and anticarcinogenic properties. Consumption of green tea polyphenols (epicatechin, epicathechin-3-gallate, epigallocetechin, and epigallocatechin-3-gallate) decreases UVR-induced erythema.

Furthermore, topical tea polyphenols are more potent than both vitamins C and E in scavenging ROS; however, they have low stability and a short duration of biologic activity. Interestingly, studies have also shown that human consumption of chocolate rich in flavanols can provide protection from UVR (Afaq and Katiyar, 2011).

Other Antioxidants

Free radicals damage DNA, lipid membranes, protein structures, and contribute to photoaging, therefore many cosmeceutical companies have tried incorporating high concentrations of antioxidants into products to improve skin care. Vitamin C is a water soluble antioxidant (AO) that neutralizes free radicals, increases collagen synthesis, and reduces collagenase expression. Topical application of vitamin C has photoprotective effects inducing reducing erythema and sunburn cell formation. Vitamin E is a lipid-soluble AO, with the most abundant form being α-tocopherol. In contrast to vitamin C, vitamin E readily reaches the stratum corneum, helping to slow the process of collagen breakdown. Together, vitamins C and E work synergistically, with vitamin C regenerating oxidized vitamin E (Chen, Hu, and Wang, 2012).

Other important antioxidants include selenium, silymarin, and soy isoflavones. Selenium helps optimize glutathione peroxidase and thioredoxin reductase and serves as a cofactor of vitamin E regeneration. The form L-selenomethionine has superior transepidermal delivery, and when combined with vitamin E, can reduce UVR induced blistering, pigmentation, and skin tumors (Chen et al., 2012).

Silymarin, from milk thistle, contains three flavonoids that have potent ROS scavenging ability and prevent lipoprotein oxidation. Finally, soybeans contain isoflavones, which have been shown to be anticarcinogenic by scavenging peroxyl and lipid radicals (Jansen et al., 2013). Other antioxidants are listed in Table 4.

Conclusion

As a skin cancer prevention strategy, there are many methods of photoprotection to consider. While much progress has been made since the first commercial sunscreen was introduced in 1928, there are many other agents, including systemic agents, which warrant further investigation.

Currently, topical protective agents remain the mainstay of photoprotection in the United States, but ineffective application and the need for reapplication often limit their effectiveness. Avoidance of sunlight during peak hours, the use of improved architectural and automobile structures, and the incorporation of photoprotective clothing, hats, and sunglasses are all important to reduce UVR exposure and enhance protection of the skin.

Table 4. Antioxidants and their functions

Antioxidant Compound	Sources	Functions
Vitamin A (retinols, carotenoids)	Colored fruits and vegetables	Increased collagen production to reduce photoaging, systemically can be used as a preventative measure for skin cancers in susceptible populations
Vitamin C	Fruits, Vegetables	Cofactor in collagen synthesis, reduces erythema and immunosuppression
Vitamin E	Oils, seeds, nuts, meats	Reduce photoaging, reduce cell membrane lipid peroxidation
Selenium	Corn, wheat soybean	Increases function of endogenous antioxidants, regenerates vitamin E
Silymarin	Milk thistle	UV filtering properties to decrease photocarcinogenesis, reduces immunosuppresion
Tea Polyphenols	Isolated from tea	Reduces erythema, can result in contact and allergic dermatitis
Soy Isoflavones (genistein, daidzein, equol)	Soy, red clover, gingko biloba	Preserve epidermal proliferation and repair mechanisms
Caffeic Acid	Coffee beans, propolis plant seeds	Shown to decrease erythema and immunosuppresion
Apigenin	Fruits and leafy vegetables, tea, wine	Decreases photoaging and photocarcinogenesis
Resveratrol	Skin and seeds of grapes, nuts, fruits	Decreases erythema and photocarcinogenesis

References

Afaq, F. and Katiyar, S. K. (2011). Polyphenols: Skin photoprotection and inhibition of photocarcinogenesis. *Mini Reviews in Medicinal Chemistry,* 11(14), 1200-1215. doi:BSP/MRMC/E-Pub/262 [pii].

Aguilera, P., Carrera, C., Puig-Butille, J. A., Badenas, C., Lecha, M., Gonzalez, S., Puig, S. (2013). Benefits of oral polypodium leucotomos extract in MM high-risk patients. *J. Eur. Acad. Dermatol. Venereol.,* 27 (9), 1095-100. doi:10.1111/j.1468-3083.2012.04659.x.

Almutawa, F. and Buabbas, H. (2014). Photoprotection: Clothing and glass. *Dermatologic Clinics,* 32(3), 439-48, x. doi:10.1016/j.det.2014.03.016 [doi].

Almutawa, F., Vandal, R., Wang, S. Q., and Lim, H. W. (2013). Current status of photoprotection by window glass, automobile glass, window films, and sunglasses. *Photodermatol. Photoimmunol. Photomed.,* 29(2), 65-72. doi: 10.1111/phpp.12022.

Bais, A. F., McKenzie, R. L., Bernhard, G., Aucamp, P. J., Ilyas, M., Madronich, S., and Tourpali, K. (2014). Ozone depletion and climate change: Impacts on UV radiation. *Photochemical and Photobiological Sciences: Official Journal of the European Photochemistry Association and the European Society for Photobiology,* 14(1), 19-52. doi:10.1039/c4pp90032d [doi].

Baron, E. D. and Suggs, A. K. (2014). Introduction to photobiology. *Dermatologic Clinics,* 32(3), 255-266. doi:10.1016/j.det.2014.03.002.

Brenner, M. and Hearing, V. J. (2008). The protective role of melanin against UV damage in human skin. *Photochemistry and Photobiology,* 84(3), 539-549. doi:10.1111/j.1751-1097.2007.00226.x [doi].

Butlers ST, F. S. (2010). Increased prevelance of left-sided skin cancers. *J. Am. Acad. Dermatol.,* 63, 1006-1010.

Chen, L., Hu, J. Y. and Wang, S. Q. (2012). The role of antioxidants in photoprotection: A critical review. *J. Am. Acad. Dermatol.,* 67(5), 1013-24. doi:10.1016/j.jaad.2012.02.009.

Council on Environmental Health, Section on, D., and Balk, S. J. (2011). Ultraviolet radiation: A hazard to children and adolescents. *Pediatrics,* 127(3), 588-597. doi:10.1542/peds.2010-3501 [doi].

Dain, S. J. (2012). Materials for occupational eye protectors. *Clin. Exp. Optom.,* 95(2), 129-39. doi:10.1111/j.1444-0938.2012.00704.x.

Dain, S. J., Ngo, T. P., Cheng, B. B., Hu, A., Teh, A. G., Tseng, J., and Vu, N. (2010). Sunglasses, the european directive and the european standard. *Ophthalmic Physiol. Opt.,* 30(3), 253-6. doi:10.1111/j.1475-1313.2010.00711.x.

Dessinioti, C., Antoniou, C., Katsambas, A., and Stratigos, A. J. (2011). Melanocortin 1 receptor variants: Functional role and pigmentary associations. *Photochemistry and Photobiology,* 87(5), 978-987. doi:10.1111/j.1751-1097.2011.00970.x [doi].

Diffey, B. L. (2002). Human exposure to solar ultraviolet radiation. *J. Cosmet. Dermatol.,* 1(3), 124-30. doi:10.1046/j.1473-2165.2002.00060.x.

Diffey, B. L. and Cheeseman, J. (1992). Sun protection with hats. *The British Journal of Dermatology,* 127(1), 10-12.

European Cosmetic Toiletry and Perfumery Association. (2011). In: Colipa Guidelines (Ed.), *In vitro photoprotection method task force: Method for in vitro determination of UVA protection*. Brussels.

Food and Drug Administration. (1999). In: Department of Health and Human Services (Ed.), *Sunscreen drug products for over-the-counter human use.* (Final Rule. ed.) Federal Register. doi:64 FR 27687, May 21, 1999.

Fourtanier, A., Moyal, D. and Seite, S. (2008). Sunscreens containing the broad-spectrum UVA absorber, mexoryl SX, prevent the cutaneous detrimental effects of UV exposure: A review of clinical study results. *Photodermatology, Photoimmunology and Photomedicine,* 24(4), 164-174. doi:10.1111/j.1600-0781.2008.00365.x [doi].

Gambichler, T., Altmeyer, P. and Hoffmann, K. (2002). Role of clothes in sun protection. *Recent Results Cancer Res.,* 160, 15-25.

Gambichler, T., Hatch, K. L., Avermaete, A., Altmeyer, P., and Hoffmann, K. (2002). Influence of wetness on the ultraviolet protection factor (UPF) of textiles: In vitro and in vivo measurements. *Photodermatol. Photoimmunol. Photomed.,* 18(1), 29-35. Retrieved from internal-pdf://0304089172/Gambichler-2002-Influence of wetness on the ul.pdf.

Gaspar, L. R. and Maia Campos, P. M. (2006). Evaluation of the photostability of different UV filter combinations in a sunscreen. *International Journal of Pharmaceutics,* 307(2), 123-128. doi:S0378-5173(05)00649-6 [pii].

Gerstenblith, M. R., Goldstein, A. M., Fargnoli, M. C., Peris, K., and Landi, M. T. (2007). Comprehensive evaluation of allele frequency differences of MC1R variants across populations. *Human Mutation,* 28(5), 495-505. doi:10.1002/humu.20476 [doi].

Green, A., Williams, G., Neale, R., Hart, V., Leslie, D., Parsons, P., Russell, A. (1999). Daily sunscreen application and betacarotene supplementation in prevention of basal-cell and squamous-cell carcinomas of the skin: A randomised controlled trial. *Lancet,* 354(9180), 723-9. doi:10.1016/S0140-6736(98)12168-2.

Hampton, P., Farr, P., Diffey, B., and Lloyd, J. (2004). Implication for photosensitive patients of ultraviolet A exposure in vehicles. *Br. J. Dermatol.,* 151, 873-876.

Harms, J., Lautenschlager, S., Minder, C. E., and Minder, E. I. (2009). An alpha-melanocyte-stimulating hormone analogue in erythropoietic protoporphyria. *N. Engl. J. Med.,* 360(3), 306-7. doi:10.1056/NEJMc0805682.

Harms, J. H., Lautenschlager, S., Minder, C. E., and Minder, E. I. (2009). Mitigating photosensitivity of erythropoietic protoporphyria patients by an agonistic analog of alpha-melanocyte stimulating hormone. *Photochem. Photobiol.,* 85(6), 1434-9. doi:10.1111/j.1751-1097.2009.00595.x.

Heinrich, U., Gartner, C., Wiebusch, M., Eichler, O., Sies, H., Tronnier, H., and Stahl, W. (2003). Supplementation with beta-carotene or a similar amount of mixed carotenoids protects humans from UV-induced erythema. *J. Nutr.,* 133(1), 98-101.

Heurung, A. R., Raju, S. I. and Warshaw, E. M. (2014). Benzophenones. *Dermatitis: Contact, Atopic, Occupational, Drug,* 25(1), 3-10. doi:10.1097/DER.0000000000000025 [doi].

Jansen, R., Osterwalder, U., Wang, S. Q., Burnett, M., and Lim, H. W. (2013). Photoprotection: Part II. sunscreen: Development, efficacy, and controversies. *J. Am. Acad. Dermatol.,* 69(6), 867 e1-14; quiz 881-2. doi:10.1016/j.jaad.2013.08.022.

Jansen, R., Wang, S. Q., Burnett, M., Osterwalder, U., and Lim, H. W. (2013). Photoprotection: Part I. photoprotection by naturally occurring, physical, and systemic agents. *J. Am. Acad. Dermatol.,* 69(6), 853 e1-12; quiz 865-6. doi:10.1016/j.jaad.2013.08.021.

Japan Cosmetic Industry Association. (1995). In: Toranomon 2-chome (Ed.), *JCIA measurement standard for UVA protection efficacy* (p 105 ed.). Minato-Ku Tokyo.

Kromann, N., Wulf, H. C., Eriksen, P., and Brodthagen, H. (1986). Relative ultraviolet spectral intensity of direct solar radiation, sky radiation and surface reflections. relative contribution of natural sources to the outdoor UV irradiation of man. *Photodermatol.,* 3(2), 73-82.

Lautenschlager, S., Wulf, H. C. and Pittelkow, M. R. (2007). Photoprotection. *Lancet,* 370(9586), 528-37. doi:10.1016/S0140-6736(07)60638-2.

Mckenzie, R. L., Weinreis, C., Johnston, P. V., Liley, B., Shiona, H., Kotkamp, M., ... Kondo, Y. (2008). Effects of urban pollution on uv spectral irradiances. *Atmospheric Chemistry and Physics Discussions, European Geosciences Union,* 8(2), 7149-7188.

Mitra, D., Luo, X., Morgan, A., Wang, J., Hoang, M. P., Lo, J., ... Fisher, D. E. (2012). An ultraviolet-radiation-independent pathway to melanoma carcinogenesis in the red hair/fair skin background. *Nature,* 491(7424), 449-453. doi:10.1038/nature11624 [doi].

Moise, A. F. and Aynsley, R. (1999). Ambient ultraviolet radiation levels in public shade settings. *Int. J. Biometeorol.,* 43(3), 128-38.

Nakagawara, V. B., Montgomery, R. W. and Marshall, J. W. (2007). *Optical radiation transmittance of aircraft windscreens and pilot vision.* Federal Aviation Administration, Washington, DC.

National Glass Association. (2014). *Types of glass.* Retrieved from http:// www.myglassclass.org/courses/Types/player.html.

Norval, M., Lucas, R. M., Cullen, A. P., de Gruijl, F. R., Longstreth, J., Takizawa, Y., and van der Leun, J. C. (2011). The human health effects of ozone depletion and interactions with climate change. *Photochemical and Photobiological Sciences: Official Journal of the European Photochemistry Association and the European Society for Photobiology,* 10(2), 199-225. doi:10.1039/c0pp90044c [doi].

Palm, M. D. and O'Donoghue, M. N. (2007). Update on photoprotection. *Dermatologic Therapy,* 20(5), 360-376. doi:DTH150 [pii].

Parisi, A. V., Meldrum, L. R., Kimlin, M. G., Wong, J. C., Aitken, J., and Mainstone, J. S. (2000). Evaluation of differences in ultraviolet exposure during weekend and weekday activities. *Physics in Medicine and Biology,* 45(8), 2253-2262.

Pasquali, E., Garcia-Borron, J. C., Fargnoli, M. C., Gandini, S., Maisonneuve, P., Bagnardi, V., ... M-SKIP Study Group. (2015). MC1R variants increased the risk of sporadic cutaneous melanoma in darker-pigmented caucasians: A pooled-analysis from the M-SKIP project. *International Journal of Cancer. Journal International Du Cancer,* 136(3), 618-631. doi:10.1002/ijc.29018 [doi].

Pinnell, S. R., Fairhurst, D., Gillies, R., Mitchnick, M. A., and Kollias, N. (2000). Microfine zinc oxide is a superior sunscreen ingredient to microfine titanium dioxide. *Dermatologic Surgery: Official Publication for American Society for Dermatologic Surgery [Et Al.],* 26(4), 309-314. doi:dsu99237 [pii].

Rees, J. L. (2000). The melanocortin 1 receptor (MC1R): More than just red hair. *Pigment Cell Research / Sponsored by the European Society for Pigment Cell Research and the International Pigment Cell Society,* 13(3), 135-140.

Reid, C., Fitzgerald, T., Fabre, A., and Kirby, B. (2013). Atypical melanocytic naevi following melanotan injection. *Irish Medical Journal,* 106(5), 148-149.

Rigel, D. S., Rigel, E. G. and Rigel, A. C. (1999). Effects of altitude and latitude on ambient UVB radiation. *Journal of the American Academy of Dermatology,* 40(1), 114-116. doi:S0190-9622(99)70542-6 [pii].

Rosenthal, F. S., Bakalian, A. E., Lou, C. Q., and Taylor, H. R. (1988). The effect of sunglasses on ocular exposure to ultraviolet radiation. *American Journal of Public Health,* 78(1), 72-74.

Sambandan, D. R. and Ratner, D. (2011). Sunscreens: An overview and update. *Journal of the American Academy of Dermatology,* 64(4), 748-758. doi:10.1016/j.jaad.2010.01.005 [doi].

Sanlorenzo, M., Vujic, I., Posch, C., Cleaver, J. E., Quaglino, P., and Ortiz-Urda, S. (2014). The risk of melanoma in pilots and cabin crew: UV measurements in flying airplanes. *JAMA Dermatology,* doi:10.1001/jamadermatol.2014.4643 [doi].

Schalka, S. and Silva dos Reis, V. (2011). Sun protection factor: Meaning and controversies. [Fator de proteção solar: significado e controvérsia] *An Bras. Dermatol.,* 86(3), 507-515.

Schlumpf, M., Cotton, B., Conscience, M., Haller, V., Steinmann, B., and Lichtensteiger, W. (2001). In vitro and in vivo estrogenicity of UV screens. *Environmental Health Perspectives,* 109(3), 239-244. doi:sc271_5_1835 [pii].

Singer, R. S., Hamilton, T. A., Voorhees, J. J., and Griffiths, C. E. (1994). Association of asymmetrical facial photodamage with automobile driving. *Arch. Dermatol.,* 130(1), 121-3.

Tanew, A., Radakovic, S., Gonzalez, S., Venturini, M., and Calzavara-Pinton, P. (2012). Oral administration of a hydrophilic extract of polypodium leucotomos for the prevention of polymorphic light eruption. *J. Am. Acad. Dermatol.,* 66(1), 58-62. doi:10.1016/j.jaad.2010.09.773.

Thieden, E., Philipsen, P. A., Heydenreich, J., and Wulf, H. C. (2004). UV radiation exposure related to age, sex, occupation, and sun behavior based on time-stamped personal dosimeter readings. *Archives of Dermatology,* 140(2), 197-203. doi:10.1001/archderm.140.2.197 [doi].

Tuchinda, C., Srivannaboon, S. and Lim, H. W. (2006). Photoprotection by window glass, automobile glass, and sunglasses. *J. Am. Acad. Dermatol.,* 54(5), 845-54. doi:10.1016/j.jaad.2005.11.1082.

Turnbull, D. J. and Parisi, A. V. (2003). Spectral UV in public shade settings. *Journal of Photochemistry and Photobiology. B, Biology,* 69(1), 13-19. doi:S1011134402003871 [pii].

US Environmental Protection Agency. (2012). *United states environmental protection agency: How the UV index is calculated.* Retrieved from http://www.epa.gov/sunwise/uvicalc.html.

Walsh, J. E. and Bergmanson, J. P. (2011). Does the eye benefit from wearing ultraviolet-blocking contact lenses? *Eye Contact Lens,* 37(4), 267-72. doi: 10.1097/ICL.0b013e3182235777.

World Health Organization. (2014). *UV index.* Retrieved from http://www.who.int/uv/intersunprogramme/activities/uv_index/en/

Wright, A. L., Hart, G. C. and Peirce, S. C. (1998). Clothing protection factor of a replica england football shirt. *Lancet,* 351(9117), 1706.

Young, A. R. (1997). Chromophores in human skin. *Phys. Med. Biol.,* 42(5), 789-802. Retrieved from internal-pdf://3829176424/Young-1997-Chromophores in human skin.pdf.

In: Sun Exposure
Editor: Collin P. Raines
ISBN: 978-1-63482-087-5

Chapter III

Sun Exposure and Protection Habits and Vitamin D Levels in Children and Adolescents with a History of Malignancy

Yael Levy-Shraga, MD and Dalit Modan-Moses*, MD
Pediatric Endocrinology Unit,
The Edmond and Lily Safra Children's Hospital,
Chaim Sheba Medical Center, Tel-Hashomer, Ramat-Gan, Israel

Abstract

Sun exposure, the main source of ultraviolet radiation exposure, is the major environmental risk factor for skin cancers, both melanoma and non-melanoma. The impact of exposure can be dramatically reduced by practicing sun protection behaviors. Still, recommendations for sunlight exposure are an area of controversy, as sun avoidance and sun protection practices may lead to inadequate vitamin D levels. This duality is particularly relevant to childhood cancer survivors, who are particularly vulnerable both to the hazardous effects of ultraviolet radiation and to deleterious skeletal and extra-skeletal effects of vitamin D deficiency.

* Correspondence to: dmodan@sheba.health.gov.il

These patients are at high risk for developing non-melanoma skin cancer and therefore are firmly advised to avoid or minimize sun exposure and adopt skin protection measures. On the other hand, several studies showed an inverse association between regional UV-B radiation exposure and cancer mortality, with activation of vitamin D-related pathways by sun exposure proposed as the likely mechanism. A protective role for vitamin D in the prevention of and recovery from malignancies has been suggested by observational studies, by clinical trials showing reduced cancer incidence with vitamin D supplementation, and by laboratory studies demonstrating that the physiologically active form of vitamin D, $1;25(OH)_2D_3$ has anti-cancerous effects. In this respect, it is worrisome that a number of studies demonstrated sub-optimal 25OHD levels in children and adolescents with a history of malignancy. Regarding sun habits of this population, there is very limited information in the literature comparing sun protection behaviors of childhood cancer survivors to individuals without a history of cancer. Three studies in adult survivors of childhood cancer showed inconsistent results. Only three previous studies investigated this issue in children and adolescents, showing that patients spent less time outside compared to controls, however adherence to sun protection recommendations was incomplete. Similar findings were recently observed by our group. We conclude that more attention should be paid to improve sun protection habits of survivors of childhood cancer throughout their lives. At the same time, since sunlight avoidance may result in vitamin D insufficiency and might have deleterious implications for bone health and possibly on cancer survival, dietary interventions to optimize intake of vitamin D may be required.

Abbreviations

25OHD - 25hydroxyvitamin D
$1;25(OH)_2D_3$ - 1,25-dihydroxyvitamin D
UVR - ultraviolet radiation
SCC - squamous cell carcinoma
BCC- basal cell carcinoma
MM - malignant melanoma
MED - minimal erythema dose
CSSS - Childhood Cancer Survivor Study
BMI - body mass index

Sun Exposure and Skin Cancer

Although skin cancer is the most preventable type of cancer, it is the most common and rapidly increasing form of cancer diagnosed in the US today - about 3.5 million basal and squamous cell skin cancers are diagnosed in the US each year [1]. Sun exposure, the main source of ultraviolet radiation (UVR) exposure, is the major environmental risk factor for skin cancers, both melanoma [2] and non-melanoma [3]. Epidemiological and laboratory data have convincingly shown that sunburns are implicated in the pathogenesis of squamous cell carcinoma (SCC) [4], basal cell carcinoma (BCC) [5, 6], and malignant melanoma (MM) [7, 8]. Chronic sun exposure is the most important cause for the formation of SCC [9], but may be less important for the development of BCC [6, 10]. Concerning MM, numerous epidemiologic investigations analyzing solar UV exposure parameters have consistently reported an association between the development of MM and short-term intense UV exposure, particularly a history of severe sunburn during childhood [8, 11]. However chronic, less intense exposure has not been found to be a risk factor for the development of MM and in fact has been found in several studies to be protective [7, 12]. A population-based study of survival from melanoma comprising 528 patients suggested that some factors associated with high levels of sun exposure, such as solar elastosis and, to a lesser extent, sunburns and intermittent sun exposure, are inversely associated with death from melanoma [12]. The apparently beneficial relationship between sun exposure and survival from melanoma could be mediated by the antiproliferative and proapoptotic effects of vitamin D, the skin synthesis of which depends on sun exposure. An alternative hypothesis is that sun exposure induces less aggressive melanoma [12].

Recommendations of Sun Protection Behaviors

The impact of UV exposure can be dramatically reduced by practicing sun protection behaviors. Therefore, the World Health Organization [13] , the American Academy of Dermatology [14] and the American Academy of Pediatrics [15] have issued guidelines advocating sun-protection measures, such as wearing sunscreen and protective clothing, hats and sunglasses, in addition to avoiding tanning and sunbathing.

In order to implement these recommendations, public health campaigns were developed to improve the knowledge of the general population regarding the role of UV radiation in the development of skin cancer. The first campaigns were established in Australia in the early 1980s [16]. The World Health Organization started a global UV project called INTERSUN [17], which aimed to encourage countries to take action to reduce UV-induced health risks. Sun protection policies were implemented in schools, such as 'SunSmart' [18] and 'living with sun' [19]. A recent study found that initiation of the Danish Sun Safety campaign in 2007 succeeded in reducing high-risk sun-tanning behavior [20]. Still, recommendations for sunlight exposure are an area of controversy [21]. It has been noted that vitamin D mediated positive effects of UV light were not adequately considered in most of these campaigns, that often propose strict 'no sun policy' without giving recommendations on how to prevent vitamin D deficiency [22]. Indeed, sun protective behaviors were found associated with lower vitamin D levels [23].

Thus, some investigators recently suggested modification of current sun guidelines, recommending "sensible" sun exposure enabling adequate vitamin D synthesis [24, 25]. Still, it is difficult to establish the balance between excessive and insufficient exposure, and liberalizing the sun safe message may offer people an excuse to over-expose. Therefore, current guidelines do not recommend intentional sun exposure to induce vitamin D production, especially in high risk groups [14, 15].

Sources of Vitamin D

Sun Exposure

The body's requirements of vitamin D are primarily obtained by skin exposure to UVR of specific wavelength, 290-315 (UVB) [26]. Synthesis of vitamin D depends on latitude, skin pigmentation, sunscreen use, and time of day of exposure. Synthesis of vitamin D is minimal during winter months north of 33° latitude in the northern hemisphere and south of 33° latitude in the southern hemisphere [27]. Maximal synthesis occurs between the hours of 10:00 AM and 4:00 PM in the spring, summer, and fall [28]. UVR skin exposure is measured as the minimal erythema dose (MED) or the amount of UVR that will cause minimal erythema of the skin [26]. The amount of UVR exposure that is equivalent to 1 MED depends on skin pigmentation, duration of exposure and geographical region. In the southern United States, 4 to 10 minutes of exposure are needed for pale skin to achieve 1 MED at noon during summer [24]. Exposure of 40% of the body to one-fourth MED results in generation of approximately 1000 IU of vitamin D, the minimum amount of daily vitamin D synthesis necessary to maintain concentration in the reference range [24]. Vitamin D synthesis is decreased in individuals who spend many hours indoors or wear sun protection when they go outside. Special attention should be given to individuals that avoid sun exposure due to cultural reasons and clothing. For example, serum 25hydroxyvitamin D (25OHD) levels have been shown to be very low in Israeli Ultra-Orthodox young adults [29]. Sun exposure and sun protection habits can be evaluated using validated questionnaires [30-34]. Sun exposure habits can be assessed by asking the respondents to indicate the average number of hours they spend outside in the summer between 10:00 AM and 4:00 PM, and how many times they had red or painful sunburns during a year. Sun protection habits can be assessed by measuring on an ordinal scale protective behaviors, such as using sunscreen, wearing a shirt with sleeves, wearing sunglasses, staying in the shade and wearing a hat. Skin color may be assessed using a question based on a 4-point Likert scale ranging from "pale or milky white" to "brown, dark brown or black." [30-32]. In addition, some studies estimated ambient UV exposure during the study period using recorded local UV indices [34].

Food

Dietary vitamin D may be beneficial to overcome lack of sunlight exposure. However, dietary sources of vitamin D are scarce and include mainly fatty fish (e.g. salmon, sardines, tuna and mackerel) [26]. Under most living conditions, only 10% of body's requirements can be obtained by diet [22]. Therefore, the intake of vitamin D depends on food fortification and supplements. All infant formulas are fortified with vitamin D. In the United States some brands of cow milk, fruit juice, yogurts, cheeses and breakfast cereals are also fortified. For those who are unable to achieve adequate amounts of vitamin D in their diet or who have vitamin D deficiency, vitamin D supplements are available in 2 forms: vitamin D2 (ergocalciferol), derived from plants, and vitamin D3 (cholecalciferol), synthesized by mammals.

Daily Vitamin D Requirements

In 2011, the Institute of Medicine revised the recommended dietary allowances (RDA) for vitamin D intake to be higher than previous recommendations and the American Academy of Pediatrics endorsed these recommendations [27]. Adequate vitamin D intake for infants younger than 1 year is 400 IU/day, for persons between ages 1 year to 70 the requirement is 600 IU/day and for those aged 70 and above 800 IU/day.

Recommendations for 25OHD Levels

Vitamin D status is most reliably determined by measuring serum 25OHD [35], however there is a debate regarding optimal cutoff values. The Lawson Wilkins Pediatric Endocrine Society [26] and the Institute of Medicine [36] set the cut-point for vitamin D deficiency at circulating concentrations of 25OHD at 20 ng/ml, based on the stipulation that this minimum level is necessary to support and maintain all the classical actions of vitamin D on bone and mineral health. However, a level of 30 ng/ml has been suggested as the lower limit of normal for 25OHD levels by the Endocrine Society guidelines [28], and many experts advocate such levels to confer the beneficial nonskeletal health effects of vitamin D [37].

Prevalence of Vitamin D Deficiency

National US data revealed increasing prevalence of vitamin D deficiency from the 1988–1994 survey to 2001–2004 [38]. Cross sectional studies of vitamin D status in adolescents have found deficiency in 17% to 47% of adolescents [39-42], with lower vitamin D concentrations in the winter [41-43]. Children and adolescents who are obese are at increased risk [40, 44], possibly because of sequestration of vitamin D in body fat. Certain medications, such as anticonvulsants, glucocorticoids, antifungals, and antiretroviral medications, increase requirements and predispose subjects to deficiency. In studies in adults, at the end of the winter, 42% of 15- to 49-year-old black girls and women throughout the United States had 25OHD levels below 20 ng/ml [45], and 32% of healthy students, physicians and residents at a Boston hospital were found to be vitamin D deficient [46].

Vitamin D Roles

Skeletal

The most widely established roles of vitamin D are related to serum calcium and phosphorus homeostasis and bone mineral accretion/mobilization. Without vitamin D, only 10 to 15% of dietary calcium and about 60% of phosphorus is absorbed [47].

The interaction of 1,25-dihydroxyvitamin D [1;25$(OH)_2D_3$] with the vitamin D receptor increases the efficiency of intestinal calcium absorption to 30 to 40% and phosphorus absorption to approximately 80% [47]. During childhood and adolescence, adequate vitamin D levels are required due to its important role in cell growth and skeletal structure and development [48]. Severe vitamin D deficiency is associated with reduced bone mass in adolescents [49, 50]. On the basis of results of a longitudinal prospective study of 6712 physically active girls aged 9 through 15 years, vitamin D intake during childhood is associated with reduced risk of stress fractures [51].

Extra-Skeletal

The discovery that most tissues and cells in the body have a vitamin D receptor, and many possess the enzymatic machinery to convert the primary circulating form of vitamin D, 25OHD, to the active form, 1;25$(OH)_2D_3$, has provided new insights into the function of this vitamin [47]. A growing body of evidence suggests that vitamin D deficiency is related to disease onset or severity of many chronic illnesses [47, 52] as well as to all-cause mortality [53, 54]. Maintenance of adequate vitamin D levels is required for regulation of several physiological functions, including regulation of immune function and regulation of cellular proliferation and differentiation [55].

Low levels of 25OHD are associated with increased risk of type 1 diabetes, multiple sclerosis and Crohn's disease [56-58]. Better intake of vitamin D was found to be associated with reduced risk of developing rheumatoid arthritis [59], osteoarthritis [60], type 2 diabetes [61], and cardiovascular disease [62]. In a recent study, avoidance of sun exposure was assessed in a cohort of 29,518 Swedish women [63]. The mortality rate amongst avoiders of sun exposure was approximately twofold higher compared with the highest sun exposure group. The authors suggested that following sun exposure advice that is very restrictive in countries with low solar intensity might be in fact harmful. In keeping with these findings, a recent British study prospectively evaluating 14,641 men and women aged 42-82 for 12-15 years demonstrated an association between baseline 25OHD levels and all-cause mortality, with a hazard ratio of 0.66 for baseline 25OHD $\geq$ 36ng/ml (90 nmol/L) [64].

Sun Exposure and Non-Skin Cancer

Numerous studies linked decreased sunlight exposure to non-skin cancer incidence or survival. This association has been reported as early as 1916, and repeatedly since (reviewed by Wacker & Holick) [25]. Existing evidence is derived from ecologic studies showing lower cancer rates in geographical areas with lower daily solar radiation penetrating the atmosphere (such as high latitude) [65], studies using estimated ground-level UVR exposure from the Total Ozone Mapping Spectrometer (TOMS) dataset of the National Aeronautics and Space Administration (NASA) [66], and studies assessing individual history of sun exposure using questionnaires [67]. Specifically, the protective effect of sun exposure has been shown for colorectal [65] and breast cancers [67, 68], several sub-types of non-Hodgkin's lymphoma (NHL) [69-72], and squamous cell lung, pleural, prostate, kidney, and bladder cancers [66].

The beneficial effects of sunlight against cancer are mainly attributed to its contribution to vitamin D synthesis, although other mechanisms have been also suggested [73].

Vitamin D and Cancer Risk, Morbidity and Mortality

Several lines of evidence suggest that vitamin D has a role in decreasing the risk for cancer. Many studies demonstrated an inverse association between vitamin D and its metabolites and cancer morbidity and mortality, particularly colorectal, breast, and prostate cancer [37, 74, 75]. Furthermore, high 25OHD and high vitamin D intake at the time of diagnosis and initiation of anti-cancer treatment were associated with improved overall and recurrence-free survival [76-80] , and a recent meta-analysis demonstrated improved overall survival as well as cancer-specific survival for breast cancer, colorectal cancer and lymphoma patients with higher vs. lower (highest quartile vs. lowest quartile) 25OHD levels. For lung cancer patients results were inconclusive [81]. Only a single study was undertaken in pediatric patients, assessing 123 children undergoing hematopoietic stem cell transplantation (HSCT), and showing a more rapid increase in neutrophil granulocytes and significantly better overall survival in vitamin D sufficient patients. Furthermore, rejection and relapse rates were lower in patients with sufficient vitamin D levels [82]. These epidemiological and observational findings are supported by clinical trials showing reduced cancer incidence [83] and improved survival [84] with vitamin D supplementation.

The theory that vitamin D can help prevent cancer is biologically plausible, as the vitamin D receptor is expressed in most tissues [47]. Studies in cell culture and experimental models suggest that 1;25$(OH)_2D_3$ (calcitriol), the active metabolite of vitamin D, promotes cell differentiation, inhibits cancer-cell proliferation and exhibits anti-inflammatory, proapoptotic, and antiangiogenic properties [37, 74, 85] – processes that may regulate cancer development and progression.

Vitamin D Levels in Pediatric Patients with Malignancy

A number of studies assessed vitamin D status in pediatric patients with malignancy. Older studies included relatively small numbers of patients, and measured vitamin D levels mostly in the context of evaluating skeletal health, while other aspects of vitamin D insufficiency were not considered. Mean 25OHD levels in all studies were in the insufficiency range [86-93].

With the emerging wealth of data regarding the association between vitamin D and cancer incidence, morbidity, and mortality, several recent studies investigated this issue more comprehensively [94-97]. In a cohort of 78 pediatric ALL survivors, 53% of participants were 25OHD insufficient (15-29 ng/dl), and 12% were deficient (<15 ng/dl). Only 27% of conventional chemotherapy-treated ALL survivors and 8% of HSCT-treated ALL survivors met RDA for dietary vitamin D intake [94]. Choudhary et al measured 25OHD levels in 481 survivors of childhood cancer and found that 29% of the participants were 25OHD insufficient (<20 ng/ml) [96]. Other studies found suboptimal 25OHD levels in 35-95% of patients [34, 95, 97, 98]. The differences in rates of 25OHD deficiency may be explained by the chosen definition of optimal levels (20 ng/ml vs. 30 ng/ml), differences in patients characteristics (patients' ages, recently diagnosed vs. long-term survivors, specific diagnosis vs. mixed population, etc.), and differences in the reported rates of vitamin D supplements usage.

Risk factors for vitamin D insufficiency in these studies included older age, non-Caucasian race, and decreased ambient ultraviolet light exposure [34, 94, 96]. Higher reported dietary vitamin D intake and use of vitamin D supplementation were associated with higher serum 25OHD in one study [94], but not in another [34]. Treatment with HSCT was associated with decreased 25OHD levels in one study [34]. No association was found between 25OHD levels and gender, body mass index (BMI) [34, 96], cancer diagnoses, presence of an endocrinopathy, exposure to radiation, or glucocorticoid medications [96].

Our group assessed vitamin D status in a cohort of 211 pediatric patients with a history of malignancy (aged 12.1±5.8y, Male=69, mean time from diagnosis 4.4±3.8y), and related 25OHD levels to lifestyle habits and to the medical history. Mean 25OHD levels in this cohort were 20.6±7.9ng/ml. Vitamin D deficiency (<15ng/ml) was found in 24.6% of the patients, and insufficiency (15-20ng/ml) in 23.2%. Younger age and amount of sun exposure were associated with higher serum 25OHD. No association was found with calcium intake, disease type, gender, BMI-standard deviation score, years since diagnosis, or undergoing HSCT. 25OHD levels during winter were significantly lower than summer levels. We considered these findings concerning, given the above discussed possible contribution of 25OHD levels to improved overall and recurrence-free survival. Furthermore, survivors of childhood cancer are at risk of secondary neoplasms [99], reduced bone mineral density [88, 99], cardiovascular disease and the metabolic syndrome [100] all of which have been associated with vitamin D deficiency/insufficiency.

Childhood Cancer Survivors and Sun Exposure

Survivors of childhood and adolescent cancer are at high risk for developing non-melanoma skin cancer. In a report from the Childhood Cancer Survivor Study (CCSS), skin cancer was the most frequently occurring subsequent cancer in the cohort, with non-melanoma skin cancer accounting for 41% of all confirmed subsequent cancers and melanoma for 3%. The incidence rate of BCC increased with age, reaching 3785.9/100,000 person-years for survivors who were 45-54 years of age. Radiation therapy, either alone or in combination with chemotherapy, was associated with an increased risk of BCC compared with no chemotherapy or radiation, with an odds ratio of 39.8 for subjects who received 35 Gy or more to the skin site vs no radiation. Results were consistent with a linear dose-response relationship, with an excess odds ratio of 1.09 per Gy. Moreover, 91% of the non-melanoma skin cancer occurred within the previous radiation fields [101, 102]. These alarming rates indicate that childhood cancer survivors with a history of radiation treatment are especially susceptible to skin cancer and may subsequently increase their risk through sun exposure.

Despite the importance of sun protection in survivors of childhood cancer, there is very limited information in the literature comparing sun protection behaviors of survivors to individuals without a history of cancer. It is also unclear if childhood cancer survivors with a history of radiation treatment utilize sun protection behaviors at equivalent or different rates than survivors without a radiation history.

Several studies assessed sun-behaviors in adult survivors of childhood cancer. In a study comprising 835 survivors from the Swiss Childhood Cancer Survivor Study (SCCSS) fewer survivors than controls reported protecting themselves from sun exposure (78% vs 87%)[103]. In contrast, in a study comprising adult survivors from the USA CCSS (9298 survivors and 2950 sibling controls), survivors and siblings showed similar patterns of sunscreen use (67% vs 66%), and survivors were significantly less likely to report having sunbathed in the previous year or use artificial tanning. Survivors with radiation exposure showed increased use of sunscreen and less sunbathing or artificial tanning compared to those who were not treated with radiation, suggesting increased awareness of the need of sun protection in this population. Significant factors for regular sunscreen use in the survivor population were being female, having lighter skin complexions, having previously been examined for skin cancer, and having skin that burned when in the sun unprotected [32]. Finally, a British study assessed sun behavior as a component of a Health Behavior Index in 178 young adult survivors of childhood cancer, finding that survivors lead a healthier lifestyle compared to controls. However, sun behavior was assessed using a single question - how many times they were sunburnt in the 12 months preceding the study, and details regarding sun behavior were not presented [104].

We identified only three studies investigating sun habits in children and adolescents with a history of malignancy, all suggesting low compliance with sun behavior recommendations. In a study comprising 75 adolescent survivors of childhood cancer, 37% were non-adherent to sun protection recommendations. No control group was included [105]. In a study comprising 95 children and adults (44 children/adolescents, 12 young adults and 39 adults) who had undergone HSCT, mean self-reported hours of sun exposure was 1.1 hours/day. The majority of participants reported "rarely" or "never" using sunscreen, although a greater percentage of young adults (50%) reported "always" using sunscreen compared to the percentages of children/adolescents (16%) or adults (21%). There were no significant differences in average daily sun exposure by sunscreen use category. Again, no control group was included [98]. Simmons et. al. assessed sun habits in 22 pediatric patients undergoing allogeneic HSCT. Frequency of sunscreen use was low for both patients and controls, with only 14% of the patients reporting "always" using sunscreen, and 50% of the patients reporting "never" using sunscreen. Time spent outside and skin exposure sores were significantly lower for patients compared to controls [34].

In a recent study we assessed sun habits in 143 children with a history of malignancy and 150 healthy controls (aged 2-21 years) using validated questionnaires. Patients and controls reported similar sun exposure time during weekdays (94±82minutes/day vs. 81±65minutes/day; p=0.83), however during weekends patients spent significantly less time outside compared to controls (103±85minutes/day vs. 124±87minutes/day; p=0.015). Time elapsed from diagnosis positively correlated with time spent outside both during weekdays (r=0.194, p=0.02) and weekends (r=0.217, p=0.009). Regarding sun protection habits, patients were more likely than controls to wear a hat when in the sun (34.5% vs. 20.7% reporting "always" or "frequently"; p=0.009). However, there was no difference between the two groups regarding the frequency of using sunscreen, wearing a shirt covering the shoulders, staying in the shade or wearing sunglasses. Age and time elapsed from diagnosis were identified as risk factors for poor adherence with sun recommendations [106].

Although this study contributes to the understanding of sun protection behaviors in survivors of childhood cancer, further research is needed to determine factors impacting the decision to engage in sun protection behaviors, such as perception of benefits and barriers to utilization, and influence of the past medical illness. These factors have not been adequately explored in the survivors population and understanding them is imperative and may assist future interventions in this area.

Conclusion - Balancing Risks and Benefits of Sun Exposure

Survivors of childhood cancer are at a high risk for vitamin D deficiency and insufficiency. This finding is worrisome, as we consider these patients to be particularly vulnerable to the adverse effects of vitamin D deficiency in relation to bone health as well as possible effects of in the context of malignancy. Monitoring vitamin D levels in this population is therefore of high importance, and since it is difficult to obtain enough vitamin D from dietary sources, vitamin D deficiency should be treated by a supplement to achieve adequate levels.

At the same time, it is important to recognize that beneficial effects of UVR exposure may not occur solely through UVR-induced vitamin D synthesis. Thus, maintaining current sun avoidance policies while supplementing with vitamin D may not be sufficient to avoid the risks of insufficient exposure to UVR [25].

Sun recommendations may need to be individualized according to season, geographical location, skin type and the medical history. In a healthy population, "sensible" sun exposure, sufficient for synthesizing the minimum necessary daily amount of vitamin D, has been suggested by some investigators [24]. Still, sun exposure around midday should be avoided during the summer in most latitudes. The face should always be protected with a hat or sunscreen since it is most sun exposed and more prone to skin damage and skin cancer from sun exposure, while providing very little vitamin D synthesis [25]. Children and adolescents with history of irradiation are a particular risk group for skin cancer and may need more strict recommendations. Since our own study, as well as other studies in pediatric cancer survivors suggest incomplete adherence with sun recommendations, more attention should be paid to improve sun protection habits of these patients throughout their lives.

More research is warranted to define optimal sun exposure and vitamin D levels as well as strategies to increase compliance with sun recommendations.

References

[1] American Cancer Society. Cancer facts & figures. Atlanta, GA: The Society. p. v.

[2] Chang C, Murzaku EC, Penn L, Abbasi NR, Davis PD, Berwick M, et al. More skin, more sun, more tan, more melanoma. *American journal of public health.* 2014;104(11):e92-9.

[3] Xiang F, Lucas R, Hales S, Neale R. Incidence of Nonmelanoma Skin Cancer in Relation to Ambient UV Radiation in White Populations, 1978-2012: Empirical Relationships. *JAMA dermatology.* 2014;150(10):1063-71.

[4] Green A, Battistutta D. Incidence and determinants of skin cancer in a high-risk Australian population. *International journal of cancer Journal international du cancer.* 1990;46(3):356-61.

[5] Neale RE, Davis M, Pandeya N, Whiteman DC, Green AC. Basal cell carcinoma on the trunk is associated with excessive sun exposure. *Journal of the American Academy of Dermatology.* 2007;56(3):380-6.

[6] Kricker A, Armstrong BK, English DR. Sun exposure and non-melanocytic skin cancer. *Cancer causes & control* : CCC. 1994;5(4):367-92.

[7] Elwood JM, Jopson J. Melanoma and sun exposure: an overview of published studies. *International journal of cancer Journal international du cancer.* 1997;73(2):198-203.

[8] Gandini S, Sera F, Cattaruzza MS, Pasquini P, Picconi O, Boyle P, et al. Meta-analysis of risk factors for cutaneous melanoma: II. Sun exposure. *European journal of cancer.* 2005;41(1):45-60.

[9] Alam M, Ratner D. Cutaneous squamous-cell carcinoma. *The New England journal of medicine.* 2001;344(13):975-83.

[10] Gallagher RP, Hill GB, Bajdik CD, Fincham S, Coldman AJ ,McLean DI, et al. Sunlight exposure, pigmentary factors, and risk of nonmelanocytic skin cancer. I. Basal cell carcinoma. *Archives of dermatology.* 1995;131(2):157-63.

[11] Osterlind A, Tucker MA, Stone BJ, Jensen OM. The Danish case-control study of cutaneous malignant melanoma. II. Importance of UV-light exposure. *International journal of cancer Journal international du cancer.* 1988;42(3):319-24.

[12] Kennedy C, Bajdik CD, Willemze R, De Gruijl FR, Bouwes Bavinck JN, Leiden Skin Cancer S. The influence of painful sunburns and lifetime sun exposure on the risk of actinic keratoses, seborrheic warts, melanocytic nevi, atypical nevi, and skin cancer. *The Journal of investigative dermatology.* 2003;120(6):1087-93.

[13] WHO ultraviolet radiation and human health .http://www.who.int /mediacentre/factsheets/fs305/en/index.html.

[14] AAD Be Sun Smart Campaign. http://www.aad.org/public/sun/smart.html.

[15] Balk SJ, Council on Environmental H, Section on D. Ultraviolet radiation: a hazard to children and adolescents. *Pediatrics*. 2011;127(3):e791-817.

[16] Greinert R, Breitbart EW, Mohar P, Volkmer B. Health initiatives for the prevention of skin cancer. *Advances in experimental medicine and biology*. 2008;624:125-36.

[17] WHO. INTERSUN. The global UV project: a guide and compendium http://www.who.int/uv/publications/en/Intersunguide.pdf2003.

[18] Dono J, Ettridge KA, Sharplin GR, Wilson CJ. The relationship between sun protection policies and practices in schools with primary-age students: the role of school demographics, policy comprehensiveness and SunSmart membership. *Health education research*. 2014;29(1):1-12.

[19] Sancho-Garnier H, Pereira B, Cesarini P. A cluster randomized trial to evaluate a health education programme "Living with Sun at School". *International journal of environmental research and public health*. 2012;9(7):2345-61.

[20] Hansen MR, Bentzen J. High-risk sun-tanning behaviour: a quantitative study in Denmark, 2008-2011. *Public health*. 2014;128(9):777-83.

[21] Macdonald HM. Contributions of sunlight and diet to vitamin D status. *Calcified tissue international*. 2013;92(2):163-76.

[22] Reichrath J. Skin cancer prevention and UV-protection: how to avoid vitamin D-deficiency? *The British journal of dermatology*. 2009;161 Suppl 3:54-60.

[23] Linos E, Keiser E, Kanzler M, Sainani KL, Lee W, Vittinghoff E, et al. Sun protective behaviors and vitamin D levels in the US population: NHANES 2003-2006. *Cancer causes & control* : CCC. 2012;23(1):133-40.

[24] Holick MF. Sunlight and vitamin D for bone health and prevention of autoimmune diseases, cancers, and cardiovascular disease. *The American journal of clinical nutrition*. 2004;80(6 Suppl):1678S-88S.

[25] Wacker M, Holick MF. Sunlight and Vitamin D: A global perspective for health. *Dermato-endocrinology*. 2013;5(1):51-108.

[26] Misra M, Pacaud D, Petryk A, Collett-Solberg PF, Kappy M, Drug, et al. Vitamin D deficiency in children and its management: review of current knowledge and recommendations. *Pediatrics*. 2008;122(2):398-417.

[27] Golden NH, Abrams SA, Committee On N, Committee On N. Optimizing bone health in children and adolescents. *Pediatrics.* 2014;134(4):e1229-43.

[28] Holick MF, Binkley NC, Bischoff-Ferrari HA, Gordon CM, Hanley DA, Heaney RP, et al. Evaluation, treatment, and prevention of vitamin D deficiency: an Endocrine Society clinical practice guideline. *The Journal of clinical endocrinology and metabolism.* 2011;96(7):1911-30.

[29] Tsur A, Metzger M, Dresner-Pollak R. Effect of different dress style on vitamin D level in healthy young Orthodox and ultra-Orthodox students in Israel. *Osteoporosis international : a journal established as result of cooperation between the European Foundation for Osteoporosis and the National Osteoporosis Foundation of the USA.* 2011;22(11):2895-8.

[30] Glanz K, Yaroch AL, Dancel M, Saraiya M, Crane LA, Buller DB, et al. Measures of sun exposure and sun protection practices for behavioral and epidemiologic research. *Archives of dermatology.* 2008;144(2):217-22.

[31] Glanz K, Schoenfeld ER, Steffen A. A randomized trial of tailored skin cancer prevention messages for adults: Project SCAPE. *American journal of public health.* 2010;100(4):735-41.

[32] Buchanan N, Leisenring W, Mitby PA, Meadows AT, Robison LL, Hudson MM, et al. Behaviors associated with ultraviolet radiation exposure in a cohort of adult survivors of childhood and adolescent cancer: a report from the Childhood Cancer Survivor Study. *Cancer.* 2009;115(18 Suppl):4374-84.

[33] Gritz ER, Tripp MK, de Moor CA, Eicher SA, Mueller NH, Spedale JH. Skin cancer prevention counseling and clinical practices of pediatricians. *Pediatric dermatology.* 2003;20(1):16-24.

[34] Simmons J, Sheedy C, Lee H, Koh S, Alvarez J, Koyama T, et al. Prevalence of 25-hydroxyvitamin D deficiency in child and adolescent patients undergoing hematopoietic cell transplantation compared to a healthy population. *Pediatric blood & cancer.* 2013;60(12):2025-30.

[35] Pearce SH, Cheetham TD. Diagnosis and management of vitamin D deficiency. *BMJ.* 2010;340:b5664.

[36] Ross AC, Manson JE, Abrams SA, Aloia JF, Brannon PM, Clinton SK, et al. The 2011 report on dietary reference intakes for calcium and vitamin D from the Institute of Medicine: what clinicians need to know. *The Journal of clinical endocrinology and metabolism.* 2011;96(1):53-8.

[37] Mullin GE, Dobs A. Vitamin d and its role in cancer and immunity: a prescription for sunlight. *Nutrition in clinical practice: official*

publication of the American Society for Parenteral and Enteral Nutrition. 2007;22(3):305-22.

[38] Ginde AA, Liu MC, Camargo CA, Jr. Demographic differences and trends of vitamin D insufficiency in the US population, 1988-2004. *Archives of internal medicine.* 2009;169(6):626-32.

[39] Lapatsanis D, Moulas A, Cholevas V, Soukakos P, Papadopoulou ZL, Challa A. Vitamin D: a necessity for children and adolescents in Greece. *Calcified tissue international.* 2005;77(6):348-55.

[40] Dong Y, Pollock N, Stallmann-Jorgensen IS, Gutin B, Lan L, Chen TC, et al. Low 25-hydroxyvitamin D levels in adolescents: race, season, adiposity, physical activity, and fitness. *Pediatrics.* 2010;125(6):1104-11.

[41] Harkness LS, Cromer BA. Vitamin D deficiency in adolescent females. *The Journal of adolescent health : official publication of the Society for Adolescent Medicine.* 2005;37(1):75.

[42] Gordon CM, DePeter KC, Feldman HA, Grace E ,Emans SJ. Prevalence of vitamin D deficiency among healthy adolescents. *Archives of pediatrics & adolescent medicine.* 2004;158(6):531-7.

[43] Looker AC, Dawson-Hughes B, Calvo MS, Gunter EW, Sahyoun NR. Serum 25-hydroxyvitamin D status of adolescents and adults in two seasonal subpopulations from NHANES III. *Bone.* 2002;30(5):771-7.

[44] Harel Z, Flanagan P, Forcier M, Harel D. Low vitamin D status among obese adolescents: prevalence and response to treatment. *The Journal of adolescent health : official publication of the Society for Adolescent Medicine.* 2011;48(5):448-52.

[45] Nesby-O'Dell S, Scanlon KS, Cogswell ME, Gillespie C, Hollis BW, Looker AC, et al. Hypovitaminosis D prevalence and determinants among African American and white women of reproductive age: third National Health and Nutrition Examination Survey, 1988-1994. *The American journal of clinical nutrition.* 2002;76(1):187-92.

[46] Tangpricha V, Pearce EN, Chen TC, Holick MF. Vitamin D insufficiency among free-living healthy young adults. *The American journal of medicine.* 2002;112(8):659-62.

[47] Holick MF. Vitamin D deficiency. *The New England journal of medicine.* 2007;357(3):266-81.

[48] Braegger C, Campoy C, Colomb V, Decsi T, Domellof M, Fewtrell M, et al. Vitamin D in the healthy European paediatric population. *Journal of pediatric gastroenterology and nutrition.* 2013;56(6):692-701.

[49] Cheng S, Tylavsky F, Kroger H, Karkkainen M, Lyytikainen A, Koistinen A, et al. Association of low 25-hydroxyvitamin D concentrations with elevated parathyroid hormone concentrations and low cortical bone density in early pubertal and prepubertal Finnish girls. *The American journal of clinical nutrition*. 2003;78(3):485-92.

[50] Cashman KD, Hill TR, Cotter AA, Boreham CA, Dubitzky W, Murray L, et al. Low vitamin D status adversely affects bone health parameters in adolescents. *The American journal of clinical nutrition*. 2008;87(4):1039-44.

[51] Sonneville KR, Gordon CM, Kocher MS, Pierce LM, Ramappa A, Field AE. Vitamin D, calcium, and dairy intakes and stress fractures among female adolescents. *Archives of pediatrics & adolescent medicine*. 2012;166(7):595-600.

[52] Holick MF. Vitamin D: a D-Lightful health perspective. *Nutrition reviews*. 2008;66(10 Suppl 2):S182-94.

[53] Saliba W, Barnett O, Rennert HS, Rennert G .The risk of all-cause mortality is inversely related to serum 25(OH)D levels. *The Journal of clinical endocrinology and metabolism*. 2012;97(8):2792-8.

[54] Moreno LA, Valtuena J, Perez-Lopez F, Gonzalez-Gross M. Health effects related to low vitamin D concentrations: beyond bone metabolism. *Annals of nutrition & metabolism*. 2011;59(1):22-7.

[55] Bikle D. Nonclassic actions of vitamin D. *The Journal of clinical endocrinology and metabolism*. 2009;94(1):26-34.

[56] Cantorna MT, Zhu Y, Froicu M, Wittke A. Vitamin D status, 1,25-dihydroxyvitamin D3, and the immune system. *The American journal of clinical nutrition*. 2004;80(6 Suppl):1717S-20S.

[57] Hypponen E, Laara E, Reunanen A, Jarvelin MR, Virtanen SM. Intake of vitamin D and risk of type 1 diabetes: a birth-cohort study. *Lancet*. 2001;358(9292):1500-3.

[58] Munger KL, Levin LI, Hollis BW, Howard NS, Ascherio A. Serum 25-hydroxyvitamin D levels and risk of multiple sclerosis. *JAMA : the journal of the American Medical Association*. 2006;296(23):2832-8.

[59] Merlino LA, Curtis J, Mikuls TR, Cerhan JR, Criswell LA, Saag KG, et al. Vitamin D intake is inversely associated with rheumatoid arthritis: results from the Iowa Women's Health Study. *Arthritis and rheumatism*. 2004;50(1):72-7.

[60] McAlindon TE, Felson DT, Zhang Y, Hannan MT, Aliabadi P, Weissman B, et al. Relation of dietary intake and serum levels of vitamin D to progression of osteoarthritis of the knee among participants

in the Framingham Study. *Annals of internal medicine.* 1996;125(5):353-9.

[61] Pittas AG ,Dawson-Hughes B, Li T, Van Dam RM, Willett WC, Manson JE, et al. Vitamin D and calcium intake in relation to type 2 diabetes in women. *Diabetes care*. 2006;29(3):650-6.

[62] Zittermann A. Vitamin D and disease prevention with special reference to cardiovascular disease. *Progress in biophysics and molecular biology*. 2006;92(1):39-48.

[63] Lindqvist PG, Epstein E, Landin-Olsson M, Ingvar C, Nielsen K, Stenbeck M, et al. Avoidance of sun exposure is a risk factor for all-cause mortality: results from the Melanoma in Southern Sweden cohort. *Journal of internal medicine*. 2014;276(1):77-86.

[64] Khaw KT, Luben R, Wareham N. Serum 25-hydroxyvitamin D, mortality, and incident cardiovascular disease, respiratory disease, cancers, and fractures: a 13-y prospective population study. *The American journal of clinical nutrition*. 2014;100(5):1361-70.

[65] Garland CF, Garland FC. Do sunlight and vitamin D reduce the likelihood of colon cancer? *International journal of epidemiology*. 1980;9(3):227-31.

[66] Lin SW, Wheeler DC, Park Y, Cahoon EK, Hollenbeck AR, Freedman DM, et al. Prospective study of ultraviolet radiation exposure and risk of cancer in the United States. *International journal of cancer Journal international du cancer*. 2012;131(6):E1015-23.

[67] Knight JA, Lesosky M, Barnett H, Raboud JM, Vieth R. Vitamin D and reduced risk of breast cancer: a population-based case-control study. *Cancer epidemiology, biomarkers & prevention : a publication of the American Association for Cancer Research, cosponsored by the American Society of Preventive Oncology*. 2007;16(3):422-9.

[68] Bidgoli SA, Azarshab H. Role of vitamin D deficiency and lack of sun exposure in the incidence of premenopausal breast cancer: a case control study in Sabzevar, Iran. Asian Pacific journal of cancer prevention : *APJCP*. 2014;15(8):3391-6.

[69] Morton LM, Slager SL, Cerhan JR, Wang SS, Vajdic CM, Skibola CF, et al. Etiologic heterogeneity among non-Hodgkin lymphoma subtypes: the InterLymph Non-Hodgkin Lymphoma Subtypes Project. *Journal of* the *National Cancer Institute Monographs*. 2014;2014(48):130-44.

[70] Slager SL, Benavente Y, Blair A, Vermeulen R, Cerhan JR, Costantini AS, et al. Medical history, lifestyle, family history, and occupational risk factors for chronic lymphocytic leukemia/small lymphocytic lymphoma:

the InterLymph Non-Hodgkin Lymphoma Subtypes Project. *Journal of the National Cancer Institute Monographs*. 2014;2014(48):41-51.

[71] Linet MS, Vajdic CM, Morton LM, de Roos AJ, Skibola CF, Boffetta P, et al. Medical history, lifestyle, family history, and occupational risk factors for follicular lymphoma: the InterLymph Non-Hodgkin Lymphoma Subtypes Project. *Journal of the National Cancer Institute Monographs*. 2014;2014(48):26-40.

[72] Cerhan JR, Kricker A, Paltiel O, Flowers CR, Wang SS, Monnereau A, et al. Medical history, lifestyle, family history, and occupational risk factors for diffuse large B-cell lymphoma: the InterLymph Non-Hodgkin Lymphoma Subtypes Project. *Journal of the National Cancer Institute Monographs*. 2014;2014(48):15-25.

[73] Lucas RM, Ponsonby AL. Considering the potential benefits as well as adverse effects of sun exposure: can all the potential benefits be provided by oral vitamin D supplementation? *Progress in biophysics and molecular biology*. 2006;92(1):140-9.

[74] Deeb KK, Trump DL, Johnson CS. Vitamin D signalling pathways in cancer: potential for anticancer therapeutics. *Nature reviews Cancer*. 2007;7(9):684-700.

[75] Boscoe FP, Schymura MJ. Solar ultraviolet-B exposure and cancer incidence and mortality in the United States, 1993-2002. *BMC cancer*. 2006;6:264.

[76] Goodwin PJ, Ennis M, Pritchard KI, Koo J, Hood N. Prognostic effects of 25-hydroxyvitamin D levels in early breast cancer. *Journal of clinical oncology : official journal of the American Society of Clinical Oncology*. 2009;27(23):3757-63.

[77] Drake MT, Maurer MJ, Link BK, Habermann TM, Ansell SM, Micallef IN, et al. Vitamin D insufficiency and prognosis in non-Hodgkin's lymphoma. *Journal of clinical oncology : official journal of the American Society of Clinical Oncology. 2010*;28(27):4191-8.

[78] Gugatschka M, Kiesler K, Obermayer-Pietsch B, Groselj-Strele A, Griesbacher A, Friedrich G. Vitamin D status is associated with disease-free survival and overall survival time in patients with squamous cell carcinoma of the upper aerodigestive tract. *European archives of oto-rhino-laryngology : official journal of the European Federation of Oto-Rhino-Laryngological Societies (EUFOS) : affiliated with the German Society for Oto-Rhino-Laryngology - Head and Neck Surgery*. 2.4-1201:(8)268;011

[79] Zgaga L, Theodoratou E, Farrington SM, Din FV, Ooi LY, Glodzik D, et al. Plasma vitamin D concentration influences survival outcome after a diagnosis of colorectal cancer. *Journal of clinical oncology : official journal of the American Society of Clinical Oncology*. 2014;32(23):2430-9.

[80] Shanafelt TD, Drake MT, Maurer MJ, Allmer C, Rabe KG, Slager SL, et al. Vitamin D insufficiency and prognosis in chronic lymphocytic leukemia. *Blood*. 2011;117(5):1492-8.

[81] Li M, Chen P, Li J, Chu R, Xie D, Wang H. Review: The Impacts of Circulating 25-Hydroxyvitamin D Levels on Cancer Patient Outcomes: A Systematic Review and Meta-Analysis. *The Journal of clinical endocrinology and metabolism. 2014*:jc20134320.

[82] Hansson ME, Norlin AC, Omazic B ,Wikstrom AC, Bergman P, Winiarski J, et al. Vitamin D levels affect outcome in pediatric hematopoietic stem cell transplantation. *Biology of blood and marrow transplantation : journal of the American Society for Blood and Marrow Transplantation*. 2014;20.43-1537:(10)

[83] Lappe JM, Travers-Gustafson D, Davies KM, Recker RR, Heaney RP. Vitamin D and calcium supplementation reduces cancer risk: results of a randomized trial. *The American journal of clinical nutrition*. 2007;85(6):1586-91.

[84] Zeichner SB, Koru-Sengul T, Shah N, Liu Q, Markward NJ, Montero AJ, et al. Improved Clinical Outcomes Associated With Vitamin D Supplementation During Adjuvant Chemotherapy in Patients With HER2 Nonmetastatic Breast Cancer. *Clinical breast cancer*. 2014.

[85] van Ginkel PR ,Yang W, Marcet MM, Chow CC, Kulkarni AD, Darjatmoko S, et al. 1 alpha-Hydroxyvitamin D2 inhibits growth of human neuroblastoma. *Journal of neuro-oncology*. 2007;85(3):255-62.

[86] Halton JM, Atkinson SA, Fraher L, Webber CE, Cockshott WP, Tam C, et al. Mineral homeostasis and bone mass at diagnosis in children with acute lymphoblastic leukemia. *The Journal of pediatrics*. 1995;126(4):557-64.

[87] Halton JM, Atkinson SA, Fraher L, Webber C, Gill GJ, Dawson S, et al. Altered mineral metabolism and bone mass in children during treatment for acute lymphoblastic leukemia. *Journal of bone and mineral research: the official journal of the American Society for Bone and Mineral Research*. 1996;11(11):1774-83.

[88] Arikoski P, Kroger H, Riikonen P, Parviainen M, Voutilainen R, Komulainen J. Disturbance in bone turnover in children with a

malignancy at completion of chemotherapy. *Medical and pediatric oncology*. 1999;33(5):455-61.

[89] Alikasifoglu A, Yetgin S, Cetin M, Tuncer M, Gumruk F, Gurgey A, et al. Bone mineral density and serum bone turnover markers in survivors of childhood acute lymphoblastic leukemia: comparison of megadose methylprednisolone and conventional-dose prednisolone treatments. *American journal of hematology*. 2005;80(2):113-8.

[90] Othman F, Guo CY ,Webber C, Atkinson SA, Barr RD. Osteopenia in survivors of Wilms tumor. *International journal of oncology*. 2002;20(4):827-33.

[91] Marinovic D, Dorgeret S, Lescoeur B, Alberti C, Noel M, Czernichow P, et al. Improvement in bone mineral density and body composition in survivors of childhood acute lymphoblastic leukemia: a 1-year prospective study. *Pediatrics*. 2005;116(1):e102-8.

[92] El-Ziny MA, Al-Tonbary YA, Salama OS, Bakr A, Al-Marsafawy H, Elsharkawy AA. Low bone mass in children with malignant lymphoma. *Pediatric hematology and oncology*. 2007;24(8):577-85.

[93] Gunes AM, Can E, Saglam H, Ilcol YO, Baytan B. Assessment of bone mineral density and risk factors in children completing treatment for acute lymphoblastic leukemia. *Journal of pediatric hematology /oncology*. 2010;32(3):e102-7.

[94] Simmons JH, Chow EJ, Koehler E, Esbenshade A, Smith LA, Sanders J, et al. Significant 25-hydroxyvitamin D deficiency in child and adolescent survivors of acute lymphoblastic leukemia: treatment with chemotherapy compared with allogeneic stem cell transplant. *Pediatric blood & cancer*. 2011;56(7):1114-9.

[95] Sinha A, Avery P, Turner S, Bailey S, Cheetham T. Vitamin D status in paediatric patients with cancer. *Pediatric blood & cancer*. 2011;57(4):594-8.

[96] Choudhary A, Chou J, Heller G, Sklar C. Prevalence of vitamin D insufficiency in survivors of childhood cancer. *Pediatric blood & cancer*. 2013;60(7):1237-9.

[97] Esbenshade AJ, Sopfe J, Zhao Z, Li Z, Campbell K, Simmons JH, et al. Screening for vitamin D insufficiency in pediatric cancer survivors. *Pediatric blood & cancer*. 2014;61(4):723-8.

[98] Robien K, Strayer LG, Majhail N, Lazovich D, Baker KS, Smith AR, et al. Vitamin D status among long-term survivors of hematopoietic cell transplantation. *Bone marrow transplantation*. 2011;46(11):1472-9.

[99] Tylavsky FA, Smith K, Surprise H, Garland S, Yan X, McCammon E, et al. Nutritional intake of long-term survivors of childhood acute lymphoblastic leukemia: evidence for bone health interventional opportunities. *Pediatric blood & cancer*. 2010;55(7):1362-9.

[100] van Waas M, Neggers SJ, Pieters R, van den Heuvel-Eibrink MM. Components of the metabolic syndrome in 500 adult long-term survivors of childhood cancer. Annals of oncology : official journal of the European Society for Medical Oncology / ESMO. 2010;21(5):1121-6.

[101] Perkins JL, Liu Y, Mitby PA, Neglia JP, Hammond S, Stovall M, et al. Nonmelanoma skin cancer in survivors of childhood and adolescent cancer: a report from the childhood cancer survivor study. *Journal of clinical oncology : official journal of the American Society of Clinical Oncology*. 2005;23(16):3733-41.

[102] Watt TC, Inskip PD, Stratton K, Smith SA, Kry SF, Sigurdson AJ, et al. Radiation-related risk of basal cell carcinoma: a report from the Childhood Cancer Survivor Study. *Journal of the National Cancer Institute*. 2012;104(16):1240-50.

[103] Rebholz CE, Rueegg CS, Michel G, Ammann RA, von der Weid NX, Kuehni CE, et al. Clustering of health behaviours in adult survivors of childhood cancer and the general population. *British journal of cancer*. 2012;107(2):234-42.

[104] Larcombe I, Mott M, Hunt L. Lifestyle behaviours of young adult survivors of childhood cancer. *British journal of cancer*. 2002;87(11):1204-9.

[105] Tercyak KP, Donze JR, Prahlad S, Mosher RB ,Shad AT. Multiple behavioral risk factors among adolescent survivors of childhood cancer in the Survivor Health and Resilience Education (SHARE) program. *Pediatric blood & cancer*. 2006;47(6):825-30.

[106] Levy-Shraga Y, Pinhas-Hamiel O, Ben-Ami M, Yeshayahu Y, Temam V, Cohen R, Modan-Moses D. Sun Protection Habits and Calcium Intake in Children with Malignancy. *Horm. Res. Paediatr* 2014;82(Suppl. 1):81.

In: Sun Exposure
Editor: Collin P. Raines
ISBN: 978-1-63482-087-5

Chapter IV

Risk Factors for Sun Exposure during Spring Break among College Students

Marvin E. Langston, MPH[1], Stephanie G. Lashway, MPH[1] and Leslie K. Dennis, MS, PhD[1,2]
[1]Division of Epidemiology and Biostatistics, Mel and Enid Zuckerman College of Public Health, University of Arizona, Tucson, AZ, US
[2]Department of Epidemiology, College of Public Health, University of Iowa, Iowa City, IA, US

Abstract

In order to look at college students' behavioral practices prior to a sunny vacation (during spring break) along with their beliefs and attitudes, we recruited sororities and fraternities in the Midwestern USA to complete a self-administered questionnaire. Sorority and fraternity students were expected to have high UVR exposure due to a strong desire to tan. The questionnaire included information on sun exposure during spring break, sun-sensitivity, and tanning attitudes and behaviors.

Analyses examined associations between potential risk factors for spending 16 or more hours in the sun during spring break using logistic regression while controlling for the clustering effects of sororities and fraternities. Students who tanned mildly were 1.6 times more likely than those with moderate or deep tans to spend 16+ hours in the sun during spring break, suggesting a strong desire to tan. Students who spent 16+ hours in the sun during spring break were more likely to have frequented tanning beds (odds ratio of 2.4 for 11+ times vs. ≤5 times) and to have used self-tanning creams (odds ratio of 2.9) between New Years and spring break. These data provide evidence that use of artificial tanning devices and self-tanning creams or sprays among college students are related to increased intermittent sun exposure (during a spring break vacation) rather than reduced exposure.

Mistaken beliefs regarding a base tan as potentially beneficial need to be addressed by excellent science examining the base tan theory and translated to the public. Replacement of tanning bed use with safer sunless tanning creams may reduce some of the harmful UVR exposures. Education alone will not be sufficient to change sun seeking behavior as was seen here and in other studies.

Keywords: Abbreviations: CI = confidence interval; OR=odds ratio

Introduction

Sun exposure appears to be the most important risk factor in the development of skin cancer, with other forms of ultraviolet radiation (UVR) also supporting such associations. Thus, UVR is the major etiologic risk factor implicated in the development of skin cancer (Gandini et al., 2005; Boniol et al., 2012; Veierod et al., 2014; Karagas et al., 2014). The mechanism of sun exposure in relation to skin cancer may work through specific ultraviolet (UV) wavelengths (van Weelden et al., 1990). UVR is composed of electromagnetic radiation at various wavelengths with differing possible implications for both melanoma and non-melanoma skin cancers. UVR from the sun is comprised of UVA (λ=320-400nm), UVB (λ=280-320nm), and UVC (λ=200-280nm). The energy that each UV type carries is inversely related to its wavelength. Evidence suggests that both UVA and UVB exposure may contribute to the development of melanoma, the most aggressive form of skin cancer, albeit through differing pathways (Zhang & Rosdahl, 2003; Young et al., 1998). UVB seems to contribute to non-melanoma skin cancers such as squamous cell carcinoma and basal cell carcinoma (Woodhead et al., 1999).

There are distinct types of sun exposure that have varying degrees of association with skin cancer types. Sun exposure as a risk factor can be classified as chronic or total sun exposure and intermittent sun exposure. Most consistently melanoma seems to be caused by intermittent periods of high sun exposure on unaccustomed skin (Armstrong & Kricker, 2001). This represents the pattern of sun exposure and not just the amount. With intermittent exposure to the sun, the skin is more vulnerable to the effects of UV radiation as exposure may result in sunburn, solar keratoses or other sun-induced skin damage (Elwood et al., 1984; Green, 1984; Holman & Armstrong, 1984; Dubin et al., 1986; Armstrong, 1988). Research on chronic sun exposure over the life course of individuals has proven less convincing for melanoma (Armstrong, 1988). The role of sun exposure for basal cell carcinoma is unclear although some have postulated intermittent sun exposure in addition to childhood sun exposure as important factors (Madan et al., 2010). However, chronic sun exposure is a major etiologic factor for squamous-cell carcinoma (Madan et al., 2010).

Other sources of UVR include artificial tanning devices. Artificial UVR tanning or indoor tanning includes the use of tanning beds, sunlamps, and UV tanning booths at home, in a salon, or commercial location. From here on we will refer to such use as tanning bed use. Adolescents and college students have been reported to have the highest rates of UVR from sun exposure and artificial UVR sources (Cokkinides et al., 2002; Magee, 2007; Dennis et al., 2009a). Intentional tanning appears to be highest in adolescents and young adults, then dropping as adults age (Dennis et al., 2009a). Several behavioral studies in adolescents and college students found strong attitudes that tanning is important (Yoo & Hur, 2014; Benmarhnia et al., 2013; Dennis et al., 2009b). Most adolescents and young adults prefer to tan via sunbathing or use of tanning beds over using sunless tanning products (Banks et al., 1992; Boldeman et al., 2003; Dennis et al., 2009b; Geller et al., 2002; Mawn & Fleischer, 1993). In the United States, nearly 30% of non-Hispanic white female high-school students used indoor tanning in 2011 and 24.9% of non-Hispanic white women (18-34) used indoor tanning in 2010 (Guy et al., 2013). In France, nearly 14% of 20-25 year old women used tanning beds (Benmarhnia et al., 2013). Such UVR exposures have led to studies attempting to understand why tanning is so important among these age groups.

The objective of this study was to examine factors related to increased amounts of sun exposure during spring break. Considerations of these factors will help inform the focus of future interventions. We specifically examined prior tanning bed use and use of sunless tanning products to see if they were related to an increase or decrease of intermittent sun exposure during vacation (spring break). We also investigated how knowledge and attitudes regarding skin cancer prevention efforts related to spring break sun exposure. The analysis was conducted among sorority and fraternity-affiliated students in the Midwest via a cross-sectional survey.

Methods

Our interest in intermittent sun exposure predictors is due to the well described risks of this factor in melanoma and skin cancers. A University in the Midwestern United States was chosen for recruitment due to the homogenous population, later life skin cancer risks, and ability to analyze recent intermittent sun exposures. The student population of the University under study was 92% non-Hispanic white, 2.2% African Americans, 0.4% American Indians, 3.7% Asians, and 2.5% identified as Latino during the period of data collection.

Students from sororities and fraternities (Greek houses) were recruited to participate in a survey and educational session on skin care. Participants completed an informed consent document with the self-administered questionnaire. During the data collection, participants were provided pizza and were given $10 re-imbursement. This project was approved by the Institutional Review Board for Human Subjects. Only Greek houses with 10 or more members were eligible for recruitment. An estimated 80% of those attending their monthly house meeting met the initial inclusion requirements with 163 students recruited.

The self-administered questionnaire included information on sun exposure during spring break, sun-sensitivity, sunburns, artificial UVR tanning, sunless tanning cream use and tanning attitudes. For these analyses the primary outcome was high sun exposure over spring break. Sun exposure was defined as the self-reported total hours spent outdoors from sunrise to sunset. High sun exposure during spring break was a reported 16 hours or more during the week compared to 15 or fewer hours in the sun. The reliability of the self-administered questionnaire on artificial UVR tanning for specific time-periods ranged in Kappa values from 0.7 to 0.9 (Dennis et al., 2008), suggesting that these students reported their artificial UVR tanning practices consistently (Gordis, 2013). The self-administered questionnaire also included several host factors important in studies of sun exposure. Ultimately we are interested in the amount of UVR that is absorbed by the skin. UV absorption is related to skin sensitivity to the sun. The amount of UVR an individual needs to obtain abnormal redness of the skin (the beginning of a sunburn) is called their minimal erythemal dose (MED). Less UVR is needed to produce abnormal skin redness among fair-skinned individuals than is required for dark-skinned individuals (Armstrong, 1988). However, time of day, season and latitude all influence the amount of UV radiation available for absorption (Holick, 2004). Tendency to sunburn was defined as the reaction of the skin when exposed to strong sunlight for 30 minutes for the first time each summer with no protection. Tanning ability was characterized as the ability to tan after repeated and prolonged sun exposure.

All analyses were conducted within SAS version 9.3 (SAS Institute Inc., 2010). Characteristics of the sample were described using descriptive statistics. Survey sampling statistical techniques were used because we first recruited from Greek houses, and then recruited subjects from those members who attended a monthly meeting. Survey sampling methods including a finite population correction (Tryfos, 1996; SAS Institute Inc., 2010) were used to describe the means and ranges for analyzed risk factors. Sampling weights were computed from the selection probabilities at each stage, and were based on both the recruitment of Greek houses and for the participation rate within each Greek house. These weights were applied even though we did not attempt to recruit all Greek houses and only recruited from students attending the monthly house meeting (Tryfos, 1996). PROC SURVEYFREQ was used to analyze the distribution of various risk factors while accounting for potential clustering by Greek house.

The Kappa statistic was used to show agreement between self-reported tanning ability and the objectively measured value using the colorimeter. PROC SURVEYLOGISTIC was used to examine the odds ratios (ORs) for 16+ hours of sun exposure during spring break (compared to <16 hours). Ninety-five percent confidence intervals (CI) are also reported around the ORs. Based on the sampling weights used, SAS estimated the total membership of the Greek Houses.

This estimate was similar to the number reported in student enrollment records. The Taylor expansion method was used to estimate the standard errors of the estimators used (Woodruff, 1971; SAS Institute Inc., 2010). The variance estimates also accounted for the clustering of the students associated with the same Greek houses.

Models run for behaviors, beliefs, and attitudes towards tanning were adjusted for tendency to sunburn as a measure of sun sensitivity. The amount of sun exposure an individual experiences and their sun sensitivity are extremely intertwined (Armstrong, 1988). Thus, lack of adjustment or adjustment for measures of sun sensitivity that are not directly related to the skin's response to the sun, such as hair and eye color, can cause bias when trying to understand the health effects of sun exposure.

Results

The majority of study participants were female (72%) and white (99%). They ranged in age from 18-23 years. Nearly half of all study participants were sophomores in college (49%), followed by freshmen (29%). About 87% had self-reported deep or moderate tanning ability. About one third (34%) of participants reported no tendency to sunburn, while 28% burned before tanning. None of the study participants had a personal history of skin cancer. A family history of skin cancer was marginally associated with sun exposure over spring break.

Over spring break many students reported engagement in outdoor activities generally indicative of prolonged periods of sun exposure. The following represents activities students participated in at least once over spring break, and many students participated in multiple activities. Some students went swimming, or lay by a pool, lake or ocean (59%). Others engaged in water or snow skiing (27%) or worked outdoors (45%). About half (51%) toured other cities or went sightseeing. Finally about a third (34%) went hiking or biking, while another third (35%) went boating, fishing, or engaged in other water sports. An overwhelming majority (79%) of students reported participation in at least one of these outdoor activities over spring break. Participation in these activities was strongly associated with 16+ hours of sun exposure over spring break. This highlights the problem surrounding outdoor recreational activities and the need for practicing good sun safety and skin protection.

Artificial Tanning

Increased tanning bed use between New Years and spring break was related to an increased risk of spending 16+ hours in the sun during spring break (Table 1). This association did not hold with tanning bed use in the prior 12 months or lifetime use. Sunless tanning spray use prior to spring break (between New Years and spring break) was shown to be related to spending 16+ hours in the sun during spring break.

Table 1. Tanning bed and sunless tanning cream use as they are related to sun exposure during spring break among 163 sorority and fraternity students in the Midwest

	Sun Exposure Hours [1]			
	16+	<16	OR	95% CI
Between New Years to Spring Break				
Tanning bed (# of uses)				
<5 times	36	40	Ref	-
6 to 10 times	18	13	1.54	(0.67-3.52)
11+ times	37	17	2.42	(1.38-4.23)
Sunless Tanning Cream (# of uses)				
None	72	63	Ref	-
1+ times	20	6	2.92	(1.51-5.63)
In the Last Year[2]…				
Tanning bed (# of uses)				
<5 times	28	25	Ref	-
6 to 10 times	10	14	0.64	(0.33-1.23)
11+ times	54	21	1.56	(0.63-3.86)

[1] Outdoors in the sun during spring break.

[2] Use in the last year was not posed for sunless tanning cream users.

Table 2. Host factors as they are related to sun exposure during spring break among 163 sorority and fraternity students in the Midwest

Host Factors	Sun Exposure Hours			
	16+	<16	OR	95% CI
Tendency to sunburn[1]				
Severe, moderate, or a mild sunburn	31	32	Ref	-
Burn then tan, or no sunburn	62	38	1.68	(1.44-1.97)
Ability to tan[2]				
Deeply, or moderately tanned	79	63	Ref	-
No or Mildly tanned	14	7	1.60	(1.03-2.48)
Skin Color				
Medium or Dark	45	33	Ref	-
Fair	48	37	0.95	(0.83-1.09)
Family history of skin cancer[3]				
No	60	48	Ref	-
Yes	30	19	1.26	(1.00-1.59)

[1] When skin is first exposed to strong sunlight.
[2] After repeated and prolonged sun exposure.
[3] Five respondents who were not sure are excluded.

Table 3. Attitudes, behaviors, beliefs and knowledge as they are related to sun exposure during spring break among 163 sorority and fraternity students in the Midwest[1]

	Sun Exposure Hours			
	16+	<16	OR	95% CI
Tanning Attitudes and Beliefs				
I don't mind burning if it helps me get a good tan				
Disagree or neutral	58	55	Ref	-
Agree	35	30	2.44	(1.75-3.39)
I try to sit in the shade, rather than in direct sunlight on a sunny summer day				
Agree or neutral	16	30	Ref	-
Disagree	76	39	3.49	(1.33-9.18)
I use sunlamps, tanning lamps or sunbeds so that I can spend more time in the sun				
No	76	67	Ref	-
Yes	17	3	5.50	(2.18-13.87)
Knowledge about Sun Safety [2]				
Ultraviolet lamps can cause some skin damage				
Disagree or neutral	10	2	Ref	-
Agree [3]	83	68	4.04	(1.57-10.37)
Sunlamps are safer than the sun				
True	10	13	Ref	-
False [3]	82	57	2.09	(0.76-5.75)
Sunburning to get a tan is not harmful				
True	89	64	Ref	-
False [3]	4	6	2.02	(1.12-3.65)

Table 3. (Continued)

	Sun Exposure Hours			
	16+	<16	OR	95% CI
The strength of the sun's rays is increased on cloudy days				
True	71	46	Ref	-
False [3]	20	23	0.70	(0.49-1.01)
Sunlamps can be dangerous				
Disagree or neutral	9	5	Ref	-
Agree [3]	84	65	1.38	(0.32-5.92)
Sunless tanning creams are safer than the sun				
False	14	10	Ref	-
True [3]	79	60	0.93	(0.34-2.55)
The strength of the sun's rays is increased at the top of mountains				
False	20	23	Ref	-
True [3]	71	46	1.80	(0.82-3.92)
Knowledge about tanning (overall score) [4]				
Less than 70%	10	8	Ref	-
70% or more	83	62	1.03	(0.62-1.73)

[1] All models adjusted for tendency to burn.

[2] Modeling the correct response.

[3] Correct response to the knowledge question.

[4] Correct responses to the knowledge questions presented in the table were summed and a percentage of the total assigned to each participant.

However, we saw an unexpected increased risk of 16+ hours in the sun among those who self-reported their tanning ability to be only mild rather than among those who reported to be moderate or deeply tanned after repeated and prolonged exposure to the sun. Melanin levels also showed that students who spent more time in the sun during spring break had darker unexposed skin (upper inner arm or wrist) and exposed skin of the forearm than those who spent less than 16 hours in the sun.

Reasons for Tanning

Several indicators of tanning attitudes and beliefs seemed to impact sun exposure over spring break. Twelve percent of participants reported use of artificial UV tanning devices so that they could spend more time in the sun, and 47% used these devices before they knew they were going to be in the sun. Those using tanning lamps or sunbeds so that they could spend more time in the sun (Table 3), were 5.5 times more likely to report 16+ hours in the sun over spring break. Several aspects of tanning attractiveness were explored here. Personal beliefs of tanning attractiveness and media perceptions of tanning attractiveness were not associated with sun exposure over spring break. However, participants that indicated parents and friends believed tanning was attractive were more likely to spend 16+ hours in the sun over spring break. Here perceptions on tanning attractiveness from parents and friends increased the risk of 16+ hours in the sun 1.80 times (95% CI of 1.13-2.86) and 1.54 times (95% CI of 1.05-2.27) respectively.

Sun Safety

Participants were asked about several sun protective behaviors and risky behaviors to understand if such behaviors impact their sun exposure. One of the protective behaviors involved asking them if they were likely to be "sitting in the shade, rather than in direct sunlight on a sunny summer day". Participants who disagreed with that statement were more likely to report 16+ hours of sun exposure over spring break than those that agreed or gave a neutral response on the 5 point Likert scale (OR=3.49, 95% CI of 1.33-9.18). These participants would therefore spend 16+ hours outdoors in the sun while also directly in the sun's harmful UV rays. Additionally those that didn't mind burning if it helped to achieve a good tan were at risk for 16 or more hours in the sun during spring break (OR=2.44, 95% CI of 1.75-3.39). Participant's knowledge of sun safety and harmful tanning practices was assessed as part of the questionnaire. Those who believed ultraviolet lamps could cause some skin damage were at an increased risk of 16+ hours in the sun over spring break (OR=4.04, 95%CI of 1.57-10.37). Participants who correctly indicated that sunburning to get a tan is harmful were also at an increased risk of 16+ hours in the sun over spring break (OR=2.02, 95% CI of 1.12-3.65). A summary score of correct knowledge questions was designed based on responses to 7 component items. Moderate knowledge about sun safety (70% or more correct answers to component questions) was not associated with sun exposure over spring break.

Discussion

Artificial Tanning

We found that increased tanning bed use and sunless tanning product use between New Years and spring break were related to an increased risk of spending 16+ hours in the sun during spring break. This is important regarding UVR behavior. It would appear that these students used tanning beds and/or sunless tanning products in an attempt to darken their skin prior to spring break. Consistent use of sunless tanning products (lifetime use) may indicate lower tanning ability among a subset with a strong desire to tan. Melanin levels, based on colorimeter data, showed that students who used both sunless tanning creams and tanning beds between New Years and spring break had the lightest skin color (both at sun exposed and unexposed body sites) regardless of the amount of time they spent outdoors in the sun during spring break. Whereas, those who frequented tanning beds most often, 11+ times between New Years and spring break, had more melanin (darker skin color) than others (both on sun exposed and unexposed body sites) regardless of their spring break sun exposure. The increased sun exposure among students who reported a lower tanning ability (mildly tan) appears to reflect a desire to tan among those with perceived lighter skin. However, the melanin data showed darker skin among those students who spent more time in the sun, suggesting the lighter skin is more perceived than real. The darker skin is likely due to tanning bed use and sunless tanning spray use prior to spring break along with the increased sun exposure during spring break. If it was solely or mostly due to sun exposure during spring break, then we would expect higher melanin rates for exposed skin but not unexposed skin.

Our finding that tanning bed use between New Years and spring break was associated with an increased risk of spending 16+ hours in the sun is consistent with other research. Tanning beds users report using tanning beds to prepare their skin for sunny holidays or for "vacation preparation" (Knight et al., 2002; Mawn et al., 1993; Dissel et al., 2009; Rhainds et al., 1999; Young et al., 1998; Schneider et al., 2013). A study in France found that 24% of the population (15-75 years) believed artificial tanning protected them from sunburn on vacation (Benmarhnia et al., 2013). This proportion drastically increased among tanning bed users (43%) and slightly decreased (21%) in people who had never used a tanning bed (Benmarhnia et al., 2013). Another study among all ages of tanning bed users found that about 29% use them for vacation preparation (Knight et al., 2002). This percentage of utilization increased to 61% among 17-30 year olds (Knight et al., 2002).

Sunless tanning products include fake tans, spray-on tans, and sunless tanning creams and lotions. They are available over the counter and in tanning salons via spray-on booths as an alternate to UVR tanning through sun exposure or tanning bed use. Such products can provide the desired tanned look without UVR exposure to the sun or tanning beds. Dihydroxyaceton (DHA) is the chemical component of sunless tanning products that darkens the skin with no known carcinogenic effects (Fu et al., 2004). DHA was show decades ago to increase sunlight tolerance among UVA sensitive patients and more recently for UV-B, suggesting regular use along with sunscreen for improved sun safety (Howe et al., 2008; Faurschou & Wulf, 2004). Sunless tanning products may prove as a healthier alternative to artificial UVR tanning, but in our study they negatively impacted sun seeking behavior similarly to artificial UVR tanning. It is unclear if this is only due to picking a population that puts importance on tanning or if other student populations who use sunless tanning products also spend more time in the sun. Additionally, since this is an observational study, it is unclear if they would have spent more time in the sun or using tanning beds if they did not use sunless tanning products. Public health messages should aim at replacement with sunless tanning, but also other methods to decrease high sun exposure following application.

The relationship between use of sunless tanning products and sun protection habits is unclear. While one study found that sunless tanning product users were more likely to use sunscreens and other sun protection measures, another found a higher rate of sunburn among users (Stryker et al.,

2007; Brooks et al., 2006). It is unclear if the sunburns reflected lighter skin, which is why these individuals were using the sunless tanning products to tan. So, while sunless tanning may be a beneficial alternative to indoor or outdoor UVR tanning, more research is needed to understand the physiologic effects of the products and attitudes and behaviors of people who use sunless tanning products.

Reasons for Tanning

Reasons for tanning vary with most tanners reporting multiple reasons for tanning. Some reasons for use of tanning beds mimic those for tanning in general, while others diverge a bit from general reasons for tanning. Much of the literature on reasons for tanning surrounds indoor tanning. Reasons reported for tanning bed use have included to feel healthy (Poorsattar & Hornung, 2007; Rhainds et al., 1999; Dissel et al., 2009), peer pressure or influence of friends (Knight et al., 2002; Dissel et al., 2009; Mawn et al., 1993; Poorsattar & Hornung, 2007), a desire for a feeling of warmth (Schneider et al., 2013) or as treatment of skin diseases (psoriasis, acne, or dermatitis) (Knight et al., 2002). The use of tanning beds to spend more time in the sun or before exposure to the sun, which we observed, may indicate a belief in the benefits of "base tanning" prior to periods of high sun exposure. Students who spent 16+ hours in the sun during their spring break vacation had more than 7 times the chance (as those with <16 hours in the sun) of receiving a sunburn. This would suggest that any perceived "base tan" they had was not protective of sunburning during their sunny vacation. If these risky tanning behaviors are influenced by the "base tanning", educational interventions may need to address the harmful risks associated with adoption of this theory, although some recent research shows this may be difficult. The Pool Cool diffusion trial among lifeguards, across 32 geographic regions in the US, combined educational and environmental strategies for their multicomponent skin cancer prevention program (Hiemstra et al., 2012). No association was seen between Pool Cool participation and "It helps to have a good base tan" (Hiemstra et al., 2012). However, lifeguards who rated "It helps to have a good base tan" higher at baseline (1 for "not at all" up to 4 for "a great deal") also rated it higher at follow-up (Hiemstra et al., 2012). They also found that "It helps to have a good base tan" at baseline was related to ethnicity (non-Caucasians disagreed with the statement) and moderate skin cancer risk (Hiemstra et al., 2012). Other studies have found similar behavioral beliefs where use of tanning beds to gain a perceived "protective" base tan or to spend more time in the sun is seen (Poorsattar & Hornung, 2007; Dennis et al., 2009b; Knight et al., 2002). Specifically, a Swedish survey of university students found that individuals who reported more indoor tanning also reported more outdoor tanning (Jerkegren et al., 1999). Alternately other studies showed tanning bed use to reduce the frequency of sunbathing (Dennis et al., 2009b; Knight et al., 2002). More work needs to be done in this area to better understand the harms of the base tanning theory and to identify the susceptible populations that have adopted these beliefs.

We found that positive parental and peer tanning behavior were related to increased time in the sun during student's spring break vacation. These findings are supported by several studies that have found that adolescents who had a parent who tanned or permitted tanning were more likely to intentionally tan (Cokkinides et al., 2002; Cokkinides et al., 2009; Hoerster et al., 2007; Lazovich et al., 2004; Mayer et al., 2011; Stryker et al., 2004). Additionally, adolescents with friends who thought a tan was attractive or tanned intentionally were also more likely to intentionally tan (Geller et al., 2002; Hoerster et al., 2007; Lazovich et al., 2004; Mayer et al., 2011; O'Riordan et al., 2006; Stryker et al., 2004; Yoo, 2009). These findings suggest that personal beliefs of attractiveness of tanning do not influence sun exposure behaviors in this population, but peer pressure and familial pressures may have an effect. The effect of peer pressure may be exaggerated due to the sample population chosen in our study. Fraternity and sorority students self-select into these Greek houses based on social and/or academic similarities, therefore they may be more prone to adhere to behaviors or beliefs of their peers more than others in this age group.

In our study, personal beliefs regarding attractiveness of a tan were not related to more time in the sun during their spring break vacation, but other studies have found important associations with appearance based motivation in adolescent and young adult populations. A study of female students at a college in Texas found that one of the main motivations for tanning in that population was appearance-based (felt a tan increased attractiveness) (Yoo & Hur 2014). Similarly, high percentages of tanning bed users of all ages report using them to "look better" or for "improved appearance" (Poorsattar & Hornung, 2007; Dissel et al., 2009; Rhainds et al., 1999; Knight et al., 2002; Mawn et al., 1993; Lazovich et al., 2004; Monfrecola et al., 2000; Dennis et al., 2009b; Stapleton et al., 2010). A study among tanning bed users of all ages found that almost 59% of them felt tanning improved their appearance (Dissel et al., 2009). Poorsattar & Hornung, (2007) found 82% of tanning bed users aged 17-30 years used them to "look good". The proportion of tanning bed users who tan because they "enjoy a tan appearance" increased among college aged users (17-23) to 92% (Knight et al., 2002). Again, across these studies, an improved appearance was reported as a reason for tanning bed use more frequently among adolescents and young adults (77% based on 5 studies) than among adults (53% based on 3 studies).

Several studies among adolescents and young adults also suggest use of tanning beds for relaxation (Poorsattar & Hornung, 2007; Lazovich et al., 2004; Knight et al., 2002; Dennis et al., 2009b; Dissel et al., 2009; Schneider et al., 2013). This has also been reported in studies that included people aged 16 to 90 years (Rhainds et al., 1999; Mawn et al., 1993). Pooling the percentages that report using tanning beds to relax across these seven studies, about 56% of adolescents and young adults stated they used tanning beds to relax compared to 32% among studies including all ages. More specifically, the highest percentage of individuals who felt tanning indoors was relaxing (~75%) was found in an adolescent population aged 14-17 years (Lazovich et al., 2004). The study of female college students in Texas reported the other main motivation for tanning in that population to be that they found it relaxing (emotion-based) (Yoo & Hur, 2014). Knight et al.'s (2002) study of college aged (17-22) tanning bed users found that about 42% used tanning beds for relaxation. Similarly, 44% of tanning bed users aged 17-30 used them to relax (Poorsattar & Hornung, 2007).

Sun Safety

Our findings suggest that knowledge of harmful tanning practices either increases or has no effect on the risk of spending more time outdoors in the sun during spring break, which was unexpected. A study of female university students (18-26) in Australia found that those who used sunless tanning were more knowledgeable about skin cancer than those who tanned outdoors (Day et al., 2013). Additionally, students who avoided tanning or applied sunless tanning products used more sun protection than those who tanned outdoors (Day et al., 2013). In France, 89.2% of people (15-75) were aware that tanning bed use could possibly cause cancer, and that tanning bed users were slightly less likely to believe the risk of cancer (Benmarhnia et al., 2013). However, our findings were consistent with another study among American female college students who tanned indoors, which showed that tanning bed users who used the devices for appearance-enhancement were more knowledgeable about both the health and appearance-damage risks of tanning behaviors (Stapleton et al., 2010). The students in the study who were not knowledgeable about health or appearance-damage risks did not tan specifically for appearance or relaxation (Stapleton et al., 2010). In an American population, risk reduction strategies aiming solely for increased knowledge of sun safety may not be the most effective as knowledge of correct behavior here does not translate into future sun protection.

Public health efforts need to be engaged in understanding and designing successful interventions for risk reduction. Interventions with proven efficacy should be conducted in the most high risk populations. This process will take further work at understanding basic attitudes and behavioral influences on sun exposure and tanning.

Strengths and Limitations

Survey sampling analyses allowed for control of clustering by sex and fraternity/sorority as students within Greek houses tend to cluster on unknown factors. Our study population was chosen due to their high interest in tanning and higher disposable income which could allow them more options than other student populations. Survey data were collected soon after spring break to reduce problems in recalling spring break activities. Survey items were shown to be reliable in this population (Dennis et al., 2008). Studying students in the Midwest directly after spring vacation, allows us to analyze intermittent sun seeking behaviors in a population without sun exposure prior to their vacation break. However, due to recruitment of students within sororities and fraternities, the sample is not representative of all students and may not be generalizable to all student populations.

Conclusion

College fraternity and sorority students at a Midwestern University appeared to tan using tanning beds and/or sunless tanning products before spring break in order to spend more time in the sun. Mistaken beliefs regarding a base tan as potentially beneficial need to be addressed by excellent science examining the base tan theory and translated to the public. Replacement of tanning bed use with safer sunless tanning creams may reduce some of the harmful UVR exposures. Education alone will not be sufficient to change sun seeking behavior as was seen here and in other studies. As the rates of UVR related skin cancers rise, the efficacy of multipronged intervention efforts that include education, adolescent tanning bed bans, and communication of personal UVR risks should be investigated for future public health dissemination.

Acknowledgments

This research was supported in part by the National Cancer Institute, grant number R03CA099520 and K07CA104556.

References

Armstrong, BK; Kricker, A. The epidemiology of UV induced skin cancer. *J Photochem Photobiol B,* 2001;63(1):8-18.

Armstrong, BK. Epidemiology of malignant melanoma: intermittent or total accumulated exposure to the sun? *J Dermatol Surg Oncol*, 1988;14(8): 835-49.

Banks, BA; Silverman, RA; Schwartz, RH; Tunnessen, WW. Attitudes of teenagers toward sun exposure and sunscreen use. *Pediatrics,* 1992; 89(1):40–42.

Benmarhnia, T; Leon, C; Beck, F. Exposure to indoor tanning in France: a population based study. *BMC Dermatol*. 2013;13:6.

Boldeman, C; Jansson, B; Dal, H; Ullen, H. Sunbed use among Swedish adolescents in the 1990s: a decline with an unchanged relationship to health risk behaviors. *Scand J Public Health,* 2003;31(3):233–237.

Boniol, M; Autier, P; Boyle, P; Gandini, S. Cutaneous melanoma attributable to sunbed use: systematic review and meta-analysis. *BMJ,* 2012;345:e 4757.

Brooks, K; Brooks, D; Dajani, Z; Swetter, SM; Powers, E; Pagoto, S; Geller, AC. Use of artificial tanning products among young adults. *J Am Acad Dermatol*, 2006;54(6):1060–6.

Cokkinides, VE; Weinstock, MA; O'Connell, MC; Thun, MJ. Use of indoor tanning sunlamps by US youth, ages 11–18 years, and by their parent or guardian caregivers: prevalence and correlates. *Pediatrics,* 2002;109(6):1124–1130.

Cokkinides V; Weinstock, M; Lazovich, D; Ward, E; Thun, M . Indoor tanning use among adolescents in the US, 1998 to 2004. *Cancer* 2009;115:190-8.

Day, AK; Oxlad, M; Roberts, RM. Predictors of Sun-Related Behaviors Among Young Women: Comparisons Between Outdoor Tanners, Fake Tanners, and Tan Avoiders. *J Am Coll Health*, 2013; 61(6):315-322.

Dennis, LK; Kim, Y; Lowe, JB. Consistency of reported tanning behaviors and sunburn history among sorority and fraternity students. *Photodermatol Photoimmunol Photomed,* 2008;24(4): 191-198

Dennis, LK; Kancherla, V; Snetselaar, LG. Adolescent attitudes towards tanning: does age matter? *Pediatric Health*, 2009a; 3(6):565-78.

Dennis, LK; Lowe, JB; Snetselaar, LG. Tanning behavior among young frequent tanners is related to attitudes and not lack of knowledge about the dangers. *Health Educ J,* 2009b 68(3):232-243.

Dissel, M; Rotterdam, S; Altmeyer, P; Gambichler, T. Indoor tanning in North Rhine-Westphalia Germany: a self-reported survey. *Photodermatol. Photoimmunol Photomed, 2009;*25(2):94–100.

Dubin, N; Moseson, M; Pasternack, BS. Epidemiology of malignant melanoma: pigmentary traits, ultraviolet radiation, and the identification of high-risk populations. *Recent Results Cancer Res*, 1986;102:56-75.

Elwood, JM; Gallagher, RP; Hill, GB; Spinelli, JJ; Pearson, JCG; Threlfall, W. Pigmentation and skin reaction to sun as risk factors for cutaneous melanoma: Western Canada Melanoma Study. *BMJ*, 1984:99-102.

Faurschou, A; Wulf, HC. Durability of the sun protection factor provided by dihydroxyacetone. *Photodermatol Photoimmunol Photomed,* 2004;20(5):239-42.

Fu, JM; Dusza, SW; Halpern, AC. Sunless tanning. *J Am Acad Dermatol*, 2004;50(5):706-713.

Gandini, S; Sera, F; Cattaruzza, MS; Pasquini, P; Picconi, O; Boyle, P; Melchi, CF. Meta-analysis of risk factors for cutaneous melanoma: II. Sun Exposure. *Eur J Cancer,* 2005b;41(1):45-60.

Geller, AC; Colditz, G; Oliveria, S; Emmons, K; Jorgensen, C; Aweh, GN; Frazier, AL.Use of sunscreen, sunburning rates, and tanning bed use among more than 10,000 US children and adolescents. *Pediatrics,* 2002*;*109(6):1009–1014.

Gordis, L. *Epidemiology*. 5th ed. Philadelphia, PA: Saunders; 2013.

Green, A. Sun exposure and the risk of melanoma. *Australas J Dermatol*, 1984:99-102.

Guy, GP Jr; Berkowitz, Z; Watson, M; Holman, DM; Richardson, LC. Indoor Tanning Among Young Non-Hispanic White Females. *JAMA Internal Med*, 2013; 173(20):1920-1922.

Hiemstra, M; Glanz, K; Nehl, E. Changes in sunburn and tanning attitudes among lifeguards over a summer season. *J Am Acad Dermatol.* 2012 Mar;66(3):430-7.

Holick, MF. Sunlight and vitamin D for bone health and prevention of autoimmune diseases, cancers, and cardiovascular disease. *Am J Clin Nutr*, 2004;80(6 Suppl):1678S-88S.

Holman, CD; Armstrong, BK. Cutaneous malignant melanoma and indicators of total accumulated exposure to the sun: an analysis separating histogenetic types. *J Natl Cancer Inst*, 1984; 73(1):75-82.

Hoerster, KD; Mayer, JA; Woodruff, SI; Malcarne, V; Roesch, SC;Clapp, E. The influence of parents and peers on adolescent indoor tanning behavior: Findings from a multi-city sample. *J Am Acad Dermatol* 2007;57:990-7.

Howe, W; Reed, B; Dellavalle, RP. Adding over-the-counter dihydroxyacetone self-tanners to sunscreen regiments to increase ultraviolet A light protection. *J Am Acad Dermatol*, 2008;58(5):894.

Jerkegren, E; Sandrieser, L; Brandberg, Y; Rosdahl, I. Sun-related behaviour and melanoma awareness among Swedish university students. *Eur J Cancer Prev,* 1999;8(1):27-34.

Karagus, MR; Zens, MS; Li, Z; Stukel, TA; Perry, AE; Gilbert-Diamond, D; Sayarath, V; Stephenson, RS; Barton, D; Nelson, HH; Spencer, SK. Early-onset basal cell carcinoma and indoor tanning: a population-based study. *Pediatrics*, 2014;134(1):e4-12.

Knight, JM; Kirincich, AN; Farmer, ER; Hood, AF. Awareness of the risks of tanning lamps does not influence behavior among college students. *Arch Dermatol,* 2002;138(10):1311-1315.

Lazovich, D; Forster, J; Sorensen, G; Emmons, K; Stryker, J; Demierre, MF; Hickle, A; Remba, N. Characteristics associated with use or intention to use indoor tanning among adolescents. *Arch Pediatr Adolesc Med,* 2004;158(9):918–924.

Madan, V; Lear, JT; Szeimies, RM. Non-melanoma skin cancer. *Lancet*, 2010;*375*(9715): 673-685.

Magee, KH; Poorsattar, S; Seidel, KD; Hornung, RL. Tanning device usage: what are parents thinking? *Pediatr Dermatol,* 2007;24(3):216–221.

Mawn, VB; Fleischer, AB Jr. A survey of attitudes, beliefs, and behavior regarding tanning bed use, sunbathing, and sunscreen use. *J. Am. Acad. Dermatol, 1993;*29(6):959–962.

Mayer, JA; Woodruff, SI; Slymen, DJ; Sallis, JF; Forster, JL; Clapp, EJ; Gilmer, T. Adolescents' use of indoor tanning: A large-scale evaluation of psychosocial, environmental, and policy-level correlates. *Am J Public Health* 2011;101:930-8.

Monfrecola, G; Fabbrocini, G; Posteraro, G; Pini, D. What do young people think about the dangers of sunbathing, skin cancer and sunbeds? A questionnaire survey among Italians. *Photodermatol. Photoimmunol. Photomed*, 2000;16(1):15–18.

O'Riordan, DL; Field, AE; Geller, AC; Brooks, DR; Aweh, G; Colditz, GA; Frazier, AL. Frequent tanning bed use, weight concerns, and other health risk behaviors in adolescent females (United States). *Cancer Causes Control* 2006;17:679-86.

Poorsattar, SP; Hornung, RL. UV light abuse and high-risk tanning behavior among undergraduate college students. *J Am Acad Dermatol*, 2007;56(3):375-379.

Rhainds, M; De Guire, L; Claveau, J. A population-based survey on the use of artificial tanning devices in the Province of Quebec, Canada. *J. Am. Acad. Dermatol, 1999;*40(4):572–576.

SAS Institute Inc. *SAS/STAT 9.2 User's Guide*. 2nd ed. Cary, NC, SAS Institute Inc; 2010.

Schneider, S; Diehl, K; Bock, C; Schluter, M; Breitbart, E.W; Volkmer, B; Greinert, R. Sunbed use, user characteristics, and motivations for tanning: results from the German population-based SUN-study 2012. *JAMA Dermatol,* 2013;149(1):43-49.

Stapleton, J; Turrisi, R; Hillhouse, J; Robinson, JK; Abar, B. A comparison of the efficacy of an appearance-focused skin cancer intervention within indoor tanner subgroups identified by latent profile analysis. *J Behav Med,* 2010; 33:181-190.

Stryker, JE; Lazovich, D; Forster, JL; Forster, JL; Emmons, KM; Sorensen, G; Demierre, MF. Maternal/female caregiver influences on adolescent indoor tanning. *J Adolesc Health,* 2004;35:528. e1-9.

Stryker, JE; Yaroch, AL; Moser, RP; Atienza, A; Glanz, K. Prevalence of sunless tanning product use and related behaviors among adults in the United States: results from a national survey. *J Am Acad Dermatol*, 2007;56(3):387-390.

Tryfos, P. *Sampling methods for applied research : text and cases*. New York: J. Wiley;1996.

van Weelden, H; van der Putte, SC; Toonstra, J; van der Leun, JC. UVA-induced tumours in pigmented hairless mice and the carcinogenic risks of tanning with UVA. *Arch Dermatol Res*, 1990;282(5):289-94.

Veierod, MB; Couto, E; Lund, E; Adami, HO; Weiderpass, E. Host characteristics, sun exposure, indoor tanning and risk of squamous cell carcinoma of the skin. *Int J Cancer*, 2014;135(2):413-22.

Woodhead, AD; Setlow, RB; Tanaka, M. Environmental factors in nonmelanoma and melanoma skin cancer. *Journal of epidemiology/Japan Epidemiological Association*, 1999;9(6 Suppl):S102-14.

Woodruff, RS. A simple method for approximating the variance of a complicated estimate. *J Am Stat Assoc,* 1971;66: 411-414.

Yoo, JJ. Peer influence on adolescent boys' appearance management behaviors. *Adolescence,* 2009;44:1017-31.

Yoo, JJ; Hur, WM. Body-tanning attitudes among female college students. *Psychol Rep,* 2014 Apr;114(2):585-96.

Young, AR; Potten, CS; Nikaido, O; Parsons, PG; Boenders, J; Ramsden, JM; Chadwick, CA. Human melanocytes and keratinocytes exposed to UVB or UVA in vivo show comparable levels of thymine dimers. *J Invest Dermatol*, 1998;111:936-40.

Zhang, H; Rosdahl, I. Ultraviolet A and B differently induce intracellular protein expression in human skin melanocytes--a speculation of separate pathways in initiation of melanoma. *Carcinogenesis*, 2003;24:1929-34.

Index

B

C

D

H

I

J

K

L

M

N

O

P

Q

R

S

T

U

V

W

X

Y

Z